The Ocean World of Jacques Cousteau

Attack and Defense

The Ocean World of Jacques Cousteau

Volume 6

Attack and Defense

DP

THE DANBURY PRESS

*The **sheepshead,** aggressively displaying his strong teeth, does not use these as weapons of attack against other fish. Instead, it uses them to penetrate the hard shells of its sedentary prey—shelled snails, crabs, bivalves, and sea urchins.*

The Danbury Press
A Division of Grolier Enterprises Inc.

Publisher: Robert B. Clarke

Production Supervision: William Frampton

Published by Harry N. Abrams, Inc.

Published exclusively in Canada by
Prentice-Hall of Canada, Ltd.

Revised edition—1975

Project Director: Peter V. Ritner

Managing Editor: Steven Schepp
Assistant Managing Editor: Ruth Dugan
Senior Editors: Donald Dreves
              Richard Vahan
Assistant Editors: Jill Fairchild
                Sherry Knox

Creative Director and Designer: Milton Charles

Assistant to Creative Director: Gail Ash
Illustrations Editor: Howard Koslow

Production Manager: Bernard Kass

Science Consultant: Richard C. Murphy

Printed in the United States of America

4567899

LIBRARY OF CONGRESS CATALOGING
    IN PUBLICATION DATA

Cousteau, Jacques Yves.
    Attack and defense.

    (His The ocean world of Jacques Cousteau;
v. 6)
    1. Marine fauna—Behavior.   2. Animal
defenses.   3. Predation (Biology)
I. Title.
[QL122.C634 1975]   591.5'7   74-23980
ISBN 0-8109-0580-9

Contents

In the sea, as on land (in fact, wherever man has not interfered with natural processes), attack and defense keeps the life pyramid in balance. IT'S A HARD LIFE but all is not blood and gore in the struggle for existence. The equilibrium between offense and defense provides opportunities for the pleasanter sides of life: mating, childbearing, playing.

But many of the sea's creatures still possess BAD REPUTATIONS (Chapter I): the killer whale and shark because of their power and speed and carnivorous appetites, jellyfish because of their occasional stings, rays and eels because of their bizarre appearances, billfish because of their ability to fight to the finish, and other animals like the sea snakes and scorpionfish because of their potentially dangerous venoms.

Although cold-blooded marine creatures need less food than warm-blooded ones, catching a prey is very tricky in a three-dimensional world. KILLING FOR A CAUSE—HUNGER (Chapter II) is the first order of urgency in the sea as on land. In some cases, like the feeding frenzies of sharks, this killing can take the form of terrifying orgies, lasting for long periods of time with countless smaller fish consumed. But the march of the poisonous cone snail upon its hapless prey, the feathery lure of the anemones, the lovely dive-bombing of a sea gull, while less melodramatic, are just as effective.

Animals employ camouflage for both offensive and defensive purposes. COLOR ME INVISIBLE (Chapter III) is one of the most successful of military tactics. What it involves can be likened to playing with light: countershading, disruptive patterns, color changes which can occur instantly or over a period of time, false leads (like the butterflyfish's false eye near its tail), mimicry (which in the case of the sargassumfish involves growing various protuberances which help it more closely resemble a floating weed), and distracting stripes, which can be either vertical or horizontal, so long as they manage to confuse the predator.

When going into combat, primitive man probably armored himself with a few skins wrapped around his arms. Gradually evolved were bucklers and greaves and helmets, until by the dawn of the Renaissance the European man of war was entirely encased in metal from head to toe. He then shucked these shells, preferring that his engines of war largely carry the armor

while his personal movements regained some flexibility. Each one of these stages is reflected in the sea in one or more of the many animals who are LIVING IN ARMOR (Chapter IV). Some sea animals have armor that grows with them, others must change armor as they grow, and still others live in a type of borrowed armor, which they, too, must change as they increase in size.

"For he who fights and runs away/May live to fight another day." This is the idea behind STRATEGIC WITHDRAWALS (Chapter V). Some sea creatures, like scallops and nudibranchs, jet or gyrate out of the reach of powerful foes; some hide behind rocks or behind the surface (by breaking into the air flyingfish in effect "hide" behind the sea's surface); and others, like some molluscs, close up shells or dig into the ground. With these and a hundred other tricks, the creatures of the sea attempt to win another chance. Some even choose to live in exquisite castlelike limestone structures.

If battle cannot be avoided by escaping or hiding, an animal has no choice but to employ OFFENSIVE DEFENSES (Chapter VI). Almost all man's own weaponry has been inspired by nature: smoke screens, tear gas, guns, hammers, poisons, etc. Like Greek warriors insulting the enemy before fight, some animals even use sound; the great roar of an adult sea lion puts a younger rival in an uneasy frame of mind before a fight even starts. Others are less vocal and less aggressive— they simply taste terrible.

When battles do occur between members of the same species, it is almost always FIGHTING FOR TERRITORY AND SEX (Chapter VII). Given these incentives, some of the smallest and least combative creatures become as fierce as titans of the sea like the elephant seal. Most of the time, however, the throwing down of the gauntlet is enough to forestall any physical combat.

The octopus, the moray eel, and the crayfish demonstrate ANCIENT ANIMOSITIES (Chapter VIII). Probably because ages ago this group competed for the same territory or food, the trio finds itself locked by evolution into a relationship of elemental feud which lasts as long as the three species themselves endure. Other examples of "Hatfield and McCoyism" in the animal world include the dolphin and the shark; the penguin and the skua; and the beluga and the killer whale.

But all this attacking and defending ebbs at certain periods in the sea, and there is A TIME FOR PEACE.

Introduction: It's a Hard Life

"Nature red in tooth and claw"?

Civilization has the fundamental ambition of introducing a degree of order into the primeval struggle for existence. But to this day man has only succeeded in drawing a curtain between life's harsher manifestations and certain of our human sensibilities. Most of us have never visited a slaughterhouse or a fish cannery or a 24-hour-a-day illuminated chicken-and-egg farm. When we fall in love, we rarely have to dispose of our rivals by means of a billyclub. When a neighbor turns nasty and tosses beer cans into our backyard, we call the police. When the police collude with the neighbor, we call the state or federal authorities. When we play rough games, we institute rules and time periods and an umpire.

But the classic, bloodstained struggle simmers right along in easy view if we care to look for it. We need protein just as urgently as the shark needs his mouthful of red snapper even though we are fed by the farms and stockyards and trawlers of this world. The newspaper reminds us daily that lust and territorial dispute and injured vanity and greed turn into acts of savagery and murder in the best-regulated communities. At a more general level, nations hurl themselves into vast conflagrations which consume millions of men and irreplaceable treasure for causes which—in the past several hundred years at least—bear no sensible proportion to costs. Man has lots of shark in himself still, as almost all of us find out at least once or twice in our lifetime.

The creatures of the sea do not write poetry or paint pictures. Their lives are more obviously determined than ours by the basic quests of life: for survival, for food, for a mate, for a territory—for play. And in pursuit of these quests they have developed over the evolutionary eons offensive and defensive weapons in nearly every conceivable direction. Man can find a precursor of almost all his primitive or refined armament in the sea. There are animals that use the analogues of swords, spears, bows and arrows, nets, electric cattleprods, camouflage, armorplate, speed, poison—often two or more weapons in dazzling combination. Moreover, advanced animals utilize tactics and strategies. (A lioness will stampede a heard of zebra toward her invisibly crouching mate; the stratagem is roughly that employed by Napoleon at Jena.) Barracuda "herd" schools of smaller fish. Dolphins hunt in packs. A male killer whale will lure a ship away from his vulnerable family. In a properly conducted seal rookery (one not too overcrowded) there is a strictly enforced order in the access of males to females. No matter how exigent his passion, the young male must bide his time or get badly chewed up. It is all the struggle for life, for survival. With the notable exception of the explosives man concocts in his laboratories (culminating in the biggest bang of them all—the atom bomb), man can learn everything he wishes to know about the principles of attack and defense from some creature in the sea.

Still, in the sea, as elsewhere in nature where it has not been contaminated by civilized man's ecological irresponsibility, there are balances struck between attack and defense. For example, predators survive only if prey also survive. If too many sea otters devour too many urchins and abalone, the otter population soon suffers from the effects of starvation. (Then, of course, the urchin-abalone population recovers, in turn triggering a resurgence in otters. The mathematics of such particular prey-predator relationships have been worked out in research studies.) Theoretically a too-successful predator will ensure his own extinction. The thought that this might be man's destiny is intolerable, but we will need a great amount of vigilance, imagination, and sacrifice to avoid such a fate. In general the world we see and live in comprises a tapestry of delicately interlocked strands like the otter-abalone pair, rising and falling in innumerable consonant rhythms. The efficiencies of attack and defense systems have become attuned to each other over the 2 or 3 billion years that life has evolved on our planet.

It is this automatic "tuning" of the natural world, these dynamic balances between prey and predator, aggression and withdrawal, that Western civilization is disrupting—with the consequence that the job of restoring an order to nature is now man's alone. Man is an animal, and the "nature red in tooth and claw" side of his animal heritage is never far from the surface. Yet with his brain, and his languages, and his prehensile hands, man has liberated himself from most of the laws which rigorously limit the possibilities available to the rest of the animal world. In the field of weaponry he has borrowed a tiny chunk of the sun's own fire for his thermonuclear devices; armament cannot go much further than that. If he wants to, he can reduce earth to a nightmarish desert, or blow it up altogether. But why should he do these things? How much more in harmony with the other side of his animal being—his instincts for survival, for mating, childbearing, playing—he is when he turns the marvelous tools of intelligence and analysis which produced the ultimate weapon of attack to the fabrication of the ultimate defense: peace—human societies living at peace with one another, a human species living at peace with its world and with the other inhabitants of its world.

Jacques-Yves Cousteau

Chapter I. Bad Reputations

To restate an old saw—when a man bites a fish, that's good, but when a fish bites a man, that's bad. This is one way of saying it's all right if man kills an animal, but if an animal attacks man, the act is reprehensible. The animal is labeled "killer," something to be feared, hated, shunned, punished, even killed by man.

Stories of vicious attacks, true or not, give some sea creatures a bad reputation, deserved or not. When the tales are true, the animal is feared or disliked because it hurts man—it bites, stings, stabs, or infects him.

> "There are a few animals that have won themselves a bad reputation even though they have little or no effect on man. Their rating comes from man's interpretation of their attitude toward lower animals —for example, feeding in a so-called savage manner."

But in most cases a sea creature attacks only to defend itself against man's predation. Rarely does it threaten man without cause. Untrue stories about the "ferocious" behavior of a sea creature often arise when man knows little about the accused. The animal's mystery gives rise to wild tales that condemn it to man's list of "bad" creatures. But how dangerous are those sea animals with bad reputations? A few actually kill. A few maim. Some are poisonous when eaten by man. Most sting, stab, or poison and cause mild to severe discomfort to man. The poisons of sea creatures may affect either man's circulatory system or his nervous system. Those affecting the nervous system are usually the most dangerous because they may shut down the brain's signals that keep bodily functions like breathing continuing. But man is one of the larger beings that sea creatures encounter, and these poisons usually can't kill him. Very often these poisons are used defensively against predators and offensively in food gathering.

There are a few animals that have won themselves a bad reputation even though they have little or no effect on man. They have won their rating through man's interpretation of their attitude toward lower animals. These animals have been seen feeding in what appears to be a savage manner. But this behavior may perhaps be comparable to a man tearing the flesh off a chicken leg with his teeth. Some animals of the sea often seem to be feared by other sea creatures. A large predator that swims in leisurely fashion until it spots a school of small fish may provoke or seem to provoke fear in the school, and the school may suddenly and dramatically maneuver to escape. It then seems to us that the predator has a bad reputation among its fellow inhabitants of the ocean.

Killer whale. English-speaking people have dubbed the orca, largest of dolphins, the "killer whale," and until recently it has led the list of sea monsters. The reputation of these animals stems from the fact that their stomachs have been found to contain remains of walrus, seals, birds, and dolphin. Whalers have also reported that orcas drive group attacks on large whales, biting off lips and tongues; but this probably happened only to helpless harpooned whales. In fact, orcas eat almost anything, mainly fish and squid. The orca is a sleek, agile, and intelligent animal, averaging 20 to 30 feet in length. They travel in groups of 10 or 12, always led by a large male. In spite of its reputed bloodthirstiness, the orca is a rare animal, not nearly so numerous as that very successful predator, the shark. There is little doubt that the orca's powerful jaws and sharp interlocking teeth could tear a man apart, but there is no record of such an attack. There is still much to learn about orcas, but it is obvious we have misjudged them by labeling them deliberate killers.

Reef shark. *One of the group which includes some species known to attack man, this shark is common in coral reef areas of the Caribbean. Obviously, any inquisitive behavior by a shark is viewed as a threat by divers. As a consequence, reef sharks are feared, usually without reason.*

Toothy Threats

The word "shark" strikes fear into the hearts of many men. In some cases this fear is justified. In most it is not. Of the 250 species of sharks currently recognized, only about 35 to 40 are actually dangerous to humans. But, as Dr. Henry B. Bigelow of Harvard has said, it's better to get out of the water if you're uncertain. The more than 200 species of sharks not considered a peril to humans are ill equipped, physically or temperament-

The sand tiger. *With its row on row of sharp, curved teeth, the sand tiger is one of those sharks that may menace man, though not in North American waters. Sand tigers generally eat fish, crustaceans, and squid. Sand tigers are one of the few predators that attack the bluefish.*

ally, for such activity. Some have flat-topped teeth, others are too sluggish, a few are too small, and many just aren't interested in man as food or as a threat. So when they do encounter man, they go the other way—and so does man.

"Most sharks are no peril to man. When they do encounter him they go the other way—and usually so does man."

13

Stingers

Jellyfish have a bad reputation, which is deserved by some species but not by others. Pelagia have tentacles that carry poisonous stinging cells named nematocysts. So do many other jellyfish. A few, like the white moon jellies that range the oceans of the world, are harmless. Some, like the red stinging jellyfish (*Cyanea*) found in temperate waters and the sea nettle found in tropical waters and the jellyfish of the eastern United States, are usually passive but sting and release their venom if intruded upon. In man their sting causes burning and itching of the skin and at worst, swelling, redness, and breathing difficulty.

The Portuguese man-of-war secretes poison powerful enough to do serious harm to men.

Under a pink-blue gas-filled float hang many filaments reaching down 40 to 60 feet and including nematocysts, or stinging organs, capable of injecting poison. The man-of-war stings indiscriminately to feed and thus to survive. Most predators dare not touch it and only a few seabirds are known to eat it. In laboratory experiments conducted at the turn of the century, Professor Portier, working with Prince Albert I of Monaco, injected a dog with a series of decreasing doses of man-of-war venom. The dog died, and examination disclosed that it was allergic to the venom.

Pelagia. The jellyfish above have nematocysts with which they sting their prey.

Portuguese man-of-war. At right is the Portuguese man-of-war, which uses its stinging poison to stun its prey.

Manta Ray

These powerful animals have an enormous wingspan—but they are totally harmless. Still, mantas are called "devil rays" by many fishermen, and Cuban fishermen have superstitions about them (especially about their hypnotic powers and their habit of jumping out of the water onto fishing boats and performing other threatening acts). Fishermen who harpoon one of these giants soon discover its strength. A manta can demolish an ordinary fishing boat in a matter of minutes, but this is a normal fight for survival. The largest observed specimen of this animal had a wingspan of 22 feet and weighed almost two tons. The two fleshy protuberances on either side of the manta's head are believed to funnel water into its mouth, and with the water, the small fish and plankton it lives on. Far from dragging sailors to watery graves, rays are content being left alone, occasionally jumping clear of the water three times in a row.

Stingray

It is not difficult to understand why the stingray has gained its bad reputation. It has a fearsome, whiplike tail longer than its body, and near the base of this tail are one, two, or three flattened spines with small, sharp teeth—coated with venomous slime which can bring serious injury or even death to man. But our misconceptions center on the manner in which the stingray uses this formidable weapon.

A stingray leads a quiet life on the ocean's floor and never attacks man. If approached, it will flee. The stinger is used only as a defensive weapon, not as an offensive one. Its position on the tail enables the ray to sting an enemy above it. When threatened, the stingray whips its tail around until it finds its attacker. The stinger is not even used to obtain food for the ray; the ray feeds by sucking molluscs and crustaceans into its mouth. If a diver or swimmer steps on a stingray's poisonous spine, who is to blame?

Barracuda

The barracuda's razor-sharp teeth and powerful jaws coupled with its ability to strike its prey with lightning speed have given it its reputation as a killer. When it has to feed, for instance, it slashes through a school of fish with its teeth glinting, ripping, and tearing, leaving dozens of dead or dying fish in its path. It then returns to feast. It is an inquisitive animal, and its habit of hanging almost motionless in midwater watching swimmers and following their every move has furthered this reputation. The fear the barracuda causes in swimmers, divers, and fishermen, however, is not justified.

Actually this sleek powerful fish, which may grow to a length of six feet and a weight of 113 pounds, has only been involved in 30 cases of attacks. In most instances, though, the reports are unreliable. The swimmer was in turbid water, and the attack was probably accidental; the identity of the attacking fish was generally not ascertained. When a diver, even unarmed, swims toward a barracuda, the latter dashes away—but not very far—and soon it is back behind the diver, close to his swim fins. As the impression of being followed by such an enigmatic predator is somewhat disagreeable, the wary diver turns around and threatens the fish away—for a few seconds. This game of intimidation can last for hours with no apparent lassitude in the barracuda.

Barracuda. *The barracuda (above) has a reputation as a formidable predator. A streamlined shape facilitates lightning-fast attacks and knifelike teeth can sever prey in two.*

Marlin. *This fierce-looking fish (opposite) may thrash about for long periods of time at the end of a fishing line. When desperate, it may attack anything.*

Billfish

Billfish, like this small, black-striped marlin, are extremely dangerous when hooked and played on fishing tackle. This marlin has been played almost to exhaustion, and although the hook is holding fast in its mouth and the fishing line is being held taut to prevent the fish's escape, it still hasn't been landed. When nearly exhausted, marlin and related swordfish may turn and attack men, boats, or anything else near them. They have stabbed men even after being landed and while thrashing about on the deck of a fishing craft. In an underwater encounter a few years ago, a swordfish skewered the submarine *Alvin*. The fish died, *Alvin* recovered.

> "In an underwater encounter a few years ago a swordfish skewered the research submarine *Alvin*. The fish died, *Alvin* recovered."

Armless and Armed

Some of the most venomous creatures in the sea are the sea snakes. Their venom paralyzes the nervous system and the victim soon dies of suffocation. But the sea snake, including an olive brown variety found in the Great Barrier Reef off Australia, is usually unaggressive. It is often said by natives that sea snakes have a small mouth and can only bite man's tender skin at the base of the thumb. This is not true. These snakes can bite anywhere, but they only do if seriously disturbed. In the Persian Gulf many pearl divers, who did not wear goggles, have been killed by sea snakes because they could not see the snakes and accidentally grabbed them. Sea snakes range all over the tropical Pacific and Indian oceans. With the completion of a projected sea level canal through Panama, however, Pacific sea snakes may find their way into the Caribbean.

Legends tell of octopods wrapping their eight arms around men and squeezing them lifeless. In fact, however, these creatures are shy and retiring and quickly back away from an approaching diver or swimmer. But if they are attacked by their ancient enemy, the moray eel, or by man, they can and will use their strong tentacles to resist capture and may even bite. The blue-ringed octopus is a unique type of octopod. It is rarely larger than four inches long, yet in spite of its size its bite is often fatal. Beachcombers of Australia are therefore warned that these "cute" animals can be deadly, if sufficiently provoked.

Diver and octopus. The octopus (left) is not the man-crushing beast it is reputed to be and, in fact, would rather back away than fight.

Sea snake. Shy and retiring, the sea snake (above) is nonetheless one of the sea's most deadly creatures.

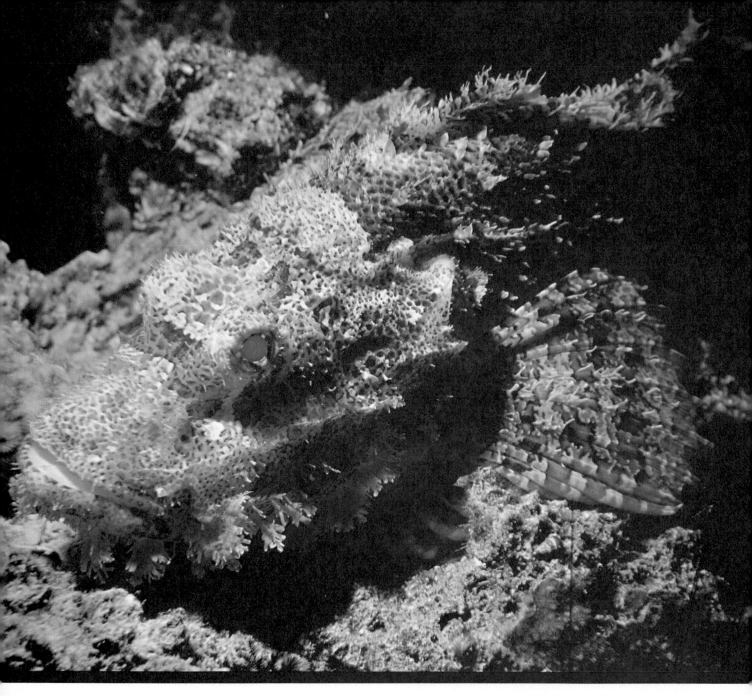

Scorpionfish

Scorpionfish are most deserving of their bad reputation. In some species their 13 dorsal spines carry the deadliest poison of any fish —poison lethal enough to kill a swimmer or beachcomber in two hours. This group includes the lionfish and the stonefish. Scorpionfish so closely resemble stones that, unfortunately, they are almost unnoticeable, and a person walking in shallow water can easily fall victim to one of them. Scorpionfish disguise is a part of their method of attack; they quickly swallow an unsuspecting smaller animal that approaches.

"Most deserving of their reputation, these fish indiscriminately sting anything that touches them. What is worse, they so closely resemble stones that they are almost unnoticeable on the bottom."

Crocodile

Looking like a fabulous creature from the Age of Reptiles, the crocodile has a fearsome reputation. These great reptiles, some of which reach more than 20 feet in length, live in tropical and subtropical fresh and salt waters. Large species, like northern Australian crocodiles, may grasp a large mammal like a cow and spin over and over, submerging the victim until it drowns. They can also stun a victim with their powerful tail. Their large teeth are one of the most powerful weapons in existence. Alligators, their relatives, which are found only in the United States and China, are less aggressive. The saltwater crocodiles of southern Asia and northern Australia, like the Nile crocodiles, are supposed to be especially vicious. In fact, crocodiles spend their lives hidden in very turbid water. On land, or in clear water, they are fearful and cautious. If man has ever been their victim, it was by accident, and long ago.

In addition to being able to attack other fish (for whatever reason) with remarkable effectiveness, the bluefish, once it has been hooked by an angler, fights furiously and jumps violently at the end of a line. When finally caught and in the boat, it has been known to grab for and bite off a fisherman's finger. Conversely, the bluefish can be made into a good meal itself once it has been caught, scaled, gutted, and broiled.

Bluefish

Bluefish are known as one of the sea's most bloodthirsty fish. These fast-moving fish do not just kill to eat—they often kill for no apparent reason. Long after their hunger is gone, they continue to slaughter, leaving shredded, half-eaten fish in their path.

School of bluefish. As they travel near shore on their annual movements up and down the east coast of the United States, bluefish (upper left) make the water boil with their activity. This fish is one of the few marine creatures that attacks when it isn't hungry.

*These **yellowtail flounders** (below and opposite) were brought to the surface shortly after a school of bluefish had traveled through, and as we can see, nips have been taken out of them. Many like them were found in the haul, and surely the ravaging was not only for the purpose of obtaining food.*

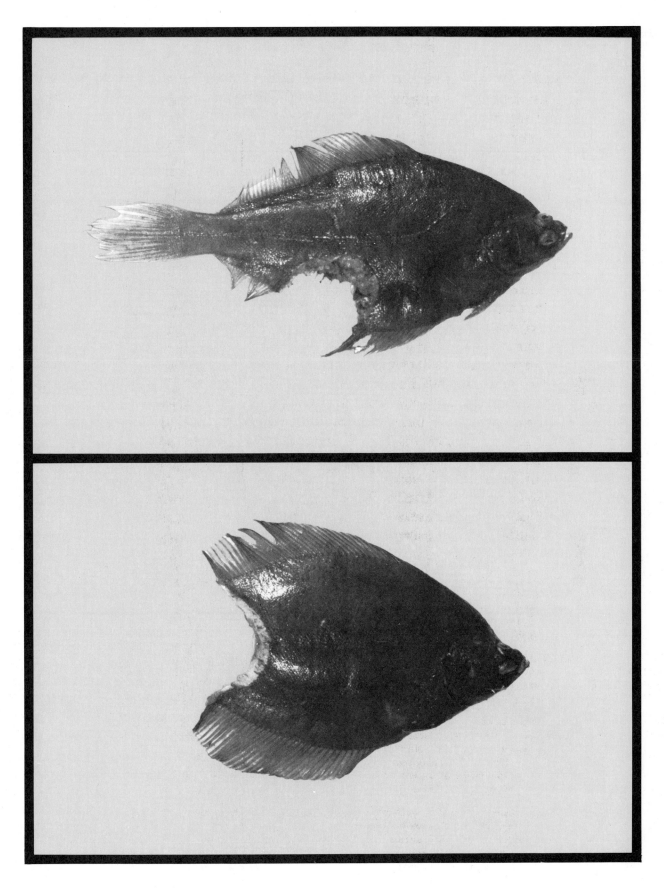

Moray Eel

Moray eels have long had a reputation for being attackers. It seems to have begun in Roman times—historians tell of Nero throwing slaves into water-filled pits of moray eels just to amuse some bored, aristocratic friends whose pleasure came from seeing people being eaten alive. In point of fact, moray eels are retiring and would rather hide than fight, and so if this tale is true, Nero must have either given the moray eels no place to escape when the slaves were cast in, or starved them until they were desperate.

These eellike fish, which are usually four to five feet long but sometimes reach ten feet, attack only when threatened. Morays threaten but won't bite unless provoked. Once they bite, however, they can do considerable injury with their strong jaws and many sharp teeth. Their manner of breathing requires them to open their mouth rhythmically every few seconds to pump water over their gills, and this action is often mistaken by divers to be a threat. Attacks on men, however, are usually the result of divers' probing into the holes in rocks and coral that shelter morays. The bite of their nasty teeth is luckily not poisonous, but wounds must be quickly treated to avoid infection. Fishermen catching a moray are advised to crush the animal's head at once.

Their snakelike, muscular bodies are perfectly suited for living in and around tropical reefs where there are nooks and crannies in which they can search for their meals and places to hide, which they do during most of the day. One of the reasons the moray and the octopus are considered enemies, in fact, is that they both seek out the same type of resting places, and all too frequently, they run into one another.

Moray eel. *This open-mouthed, fierce-looking reef fish would rather turn away than fight. Once they do feel threatened enough to fight, however, they can do great harm.*

North American Lobster

Crusher and ripper. *The lobster's claws, called the crusher and the ripper, can be formidable weapons, but the lobster will retreat if it is approached by a diver.*

It is not difficult to understand why man has developed some misconceptions about lobsters. The lobster's claws are strong, and no one who has been pinched by a lobster is likely to forget it. The big, heavy claw with blunt teeth is called the crusher claw. This is used by the lobster to crush its prey and to fight with other lobsters. The other claw, with sharp, teethlike projections, is the ripper claw and is used to tear apart food that has been crushed by the crusher claw. It is interesting to note that if a lobster is approached by a diver, it will retreat. Only if it is cornered in its hole will the lobster use its powerful claws to defend itself.

Chapter II. Killing for a Cause—Hunger

Man sometimes views the sea as a tranquil world, shielded from the violence of the land. Not so. Death is a way of life in the sea, also, and nearly all activity of cold-blooded ma-

> "Death is a way of life in the sea. Nearly all marine animal activity, both offensive and defensive, is motivated by the attempt to allay hunger or avoid predators. In the sea as on land, an appropriate motto for life is 'Eat and be eaten'."

rine creatures, both offensive and defensive, is motivated by the attempt to allay their hunger or to avoid falling prey to others. An appropriate motto for life could well be "Eat and be eaten." However, the quest for food is slightly less dramatic than on land, partly because fish need less food than birds, for example, and partly because sea mammals, which have great appetites, are so superior in intelligence to other sea creatures that they spend little time filling their stomachs.

> "Although violence and killing occur continuously in the sea, almost no animal kills unless it is feeding or fighting for survival. Prey and predator live most of the time in harmony. But whenever the predator gets hungry, the prey is gobbled up."

Hunger is one of the most powerful animal drives. To satisfy this drive, a great variety of eating habits have developed.

Tiny crustaceans and the larvae of larger animals eat microscopic plants and animals.

In turn, these diminutive hunters are preyed upon by larger hunters. And so it goes on, on up the food chain: larger animals eat smaller ones, only to be killed and consumed in their turn. And the largest creatures if unchallenged during their lives, are part of the food cycle too—they fall victim to tiny scavengers when they die.

But although violence and killing occur continuously in the sea, almost no animal kills unless it is feeding or fighting for survival. And animals usually do not kill just because food happens to be available. In fact, a fish may live safely most of the time near predators. But when the predator gets hungry and feeding conditions are right, it may be gobbled up. Hunger then is the first requisite for feeding.

The physical conditions of the oceans influence the desire to feed. These include the temperature, salinity, and acidity of the water, which vary with season and from place to place. Light plays a vital role. Deep creatures come to the surface at night, feed upon the shallow liquid meadows, and descend back to the depths at dawn, chased by daylight that they cannot endure. In the reefs, some animals hunt only under cover of darkness. But the great majority eats at dawn and dusk, during brief "feeding furors" in which birds, fish, squid, and mammals participate.

Patrolling shark. Patrolling the margins of coral reefs, sharks take advantage of the food supply found there. These magnificent creatures are usually patient hunters, waiting until an easy meal presents itself. They often feed on injured, dying, or dead animals, which have escaped other predators. These sharks show no preference for feeding either by day or by night. They often gorge themselves and then may not eat again for days.

Feeding Frenzy

Blue sharks. Attracted by the scent of food, these great fish swarm in the open ocean.

Sharks are unpredictable. Sometimes they may swim casually about a diver for hours without showing any interest in him, and on other occasions they may behave erratically, the ambient field becoming electric as soon as a diver enters the water. Sometimes sharks flee from an unarmed, unprotected swimmer, and at other times they may deliberately crash into the steel bars of anti-shark cages and bite furiously at them in unprovoked attack. The blue is usually a solitary hunter, but when the scent of blood is in the sea, many will appear as if from nowhere, like vague shadows suddenly come to life. They circle cautiously sometimes for hours until they are sure that there is no danger. Then one of the circling pack rushes the intended prey, brushing or bumping it. If the object seems edible and harmless, the boldest of the cautious group approaches for the first bite. Then a feeding frenzy may begin. Other times there may be blood in the water or a struggling wounded fish or even a shark attacking and no frenzy will occur. What initiates the shark's erratic behavior is not precisely understood. A number of factors are probably responsible for this behavior and each stimulus by itself may not be enough to initiate the frenzy. One thing, however, is sure—once a feeding frenzy begins, the behavior of those active sharks stimulates all the sharks in the vicinity to become many times more aggressive than normal. The one, odd conclusion we can come to is: the better acquainted we become with sharks, the less we know them— one can never tell what a shark is going to do.

The Shark's Eversible Jaw

Sharks *can* bite a very large object, such as the side of a whale, with their dorsal sides up. They do so by using the lower jaw and snubbing up their upper jaw in a grotesque manner. This snubbing action opens their mouths so wide that their jaws are nearly vertical, and the huge cavity of the fish's mouth is revealed, as are its rapierlike teeth. In addition, strong muscles allow the upper jaw to be thrust outward to grasp the flesh and then rotated downward in a cutting action. This efficient feeding system enables the shark to partake of a wider variety of foods than otherwise would be possible.

When it is opened wide, the shark's mouth looks like a huge steel trap ready to be sprung. When a shark takes a bite of a large animal, like a whale or dolphin, it clamps onto the animal with the great jaws and sinks its teeth into the flesh. Then it seems to go into convulsions, violently wriggling its body from head to tail. Its razor-sharp serrated teeth are twisted from side to side, and they scoop easily through the captive's flesh. This awesome spectacle is over in an instant. The bite of a shark leaves a cavity in its prey's body.

The mouths of most fish are at the extreme forward end of the body, but that of the shark is not. The shark's mouth is placed well on the underside of its head, so much so it would seem sharks would have difficulty eating anything but prey smaller than the aperture of the mouth. In fact, Aristotle thought that sharks could not bite into their food from their normal swimming position, and he postulated that they roll over to eat.

The movable jaws of some sharks enable them to bite chunks of flesh off a victim too large for one bite.

Always on the Prowl

On an expedition to the Red Sea to study and tag sharks, *Calypso* divers entered the water, protected from the unpredictable animals by antishark cages, such as the one shown above. To bring the sharks close enough for tagging, the water was churned with small bits of fish, whose scent attracted the lurking sharks. Soon sharks were all around, circling the cages. The atmosphere was tense; although the scent of food was present, there was nothing for the sharks to eat. Suddenly a diver noticed a large red snapper cruising by and decided to spear it for additional bait. But before the snapper could be retrieved by the diver, it escaped from the spear. Suddenly the spell the sharks seemed to be under was broken. A shark broke away from the pack and rushed the injured snapper, chasing it into one of the antishark cages. Then the sharks abandoned their usual caution and hurtled into the steel bars of the cage trying to get at the wounded fish. The snapper escaped from the cage but not from the sharks. In an instant they were upon it.

And what these sharks did to the snapper they *may* do to an unlucky human. At right: Arthur—a *Calypso* dummy. Equipped with backpack and fins, Arthur looked like a real diver, but during a series of tests sharks showed no interest in him until we put fish inside his wet suit. Now their interest was aroused. They began to circle him more rapidly, growing more and more excited as Arthur jiggled erratically at the end of his tether. One shark finally rushed Arthur, but veered off at the last moment. A second rush followed. Then a third shark wheeled and, with jaws agape, attacked the dummy, crunching down on one leg and tearing it off.

How Cone Snails Stab Prey

The deadly cone snail extends a hollow, venom-filled tooth from the end of its proboscis and shoots it into its quarry, where it remains. The small, barbed tooth, which looks like a miniature harpoon, is one of many radular teeth in the cone snail's radular sac. These harpoons have their origin in the filelike radular organ which is used by many snails to scrape algae off rocks. Only one tooth at a time is positioned at the tip of the proboscis, ready for use in defense or in capturing food. When a tooth is shot out, another radular tooth moves up to take its place. How the harpoon is actually shot out is not clearly understood—it may be with muscles in much the same way a slingshot is used or through a high-pressure system that rapidly forces the dart out. The venom is stored in a muscular, bulblike poison sac, which squeezes the toxin through a duct into the hollow tooth. In some species the venom is already in the tooth when it is fired, and the snail can release the tooth entirely when it has found its target. In others the snail holds on to the tooth by means of a duct through which it pumps its poison into its prey. The venom paralyzes the cone snail's victim. Then the snail moves forward to engulf its immobile prey with an extensible and fleshy mouth. Because the snail's radula cannot scrape off tissue, a large meal must be partially digested in the pharynx outside the body.

There are many types of cone snails, most of which have beautiful shells. The gloria maris is an extremely valuable cone snail, at times appraised at over one thousand dollars. Because of their beauty, collectors look for them, but a number of people have received serious stings from cone snails in the South Pacific. These collectors, unaware

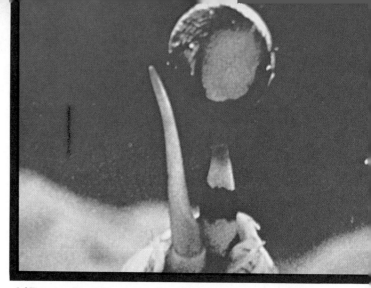

A/Preparing for attack. *The cone snail in the foreground prepares to attack a turban snail. The cone snail's proboscis is extended for the attack.*

D/Successful attack. *When the victim's skin is punctured by the cone snail's tooth, venom has already begun to spread within its victim.*

of the deadly venom apparatus, have held cone snails in their hands or put them in their pockets. This has often caused the snail to react by releasing their weapon. The poison is reported to act on the nervous system, and death in experimental animals may be from loss of respiration or heart failure.

Cone snail. *At right, an illustration of the stabbing apparatus of the cone snail, showing the poison sac, harpoonlike teeth, and the proboscis through which the hollow tooth is ejected.*

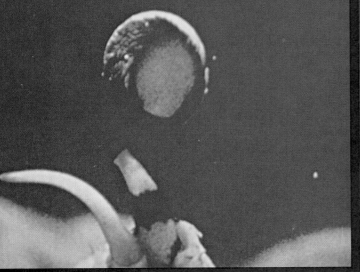
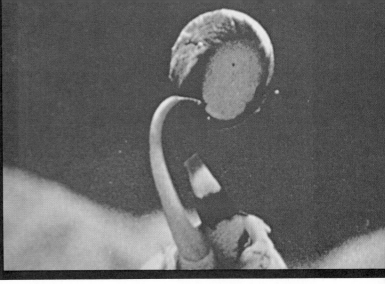

B/Proboscis in position. *As it moves to strike at the nearby snail, the cone snail waves its venomous tooth in the tip of its proboscis.*

C/Striking the victim. *The cone snail's proboscis curves around as the snail aims and lets fly its radular tooth. The victim snail is struck.*

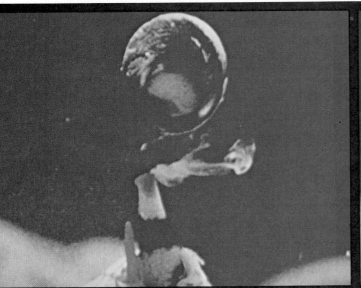
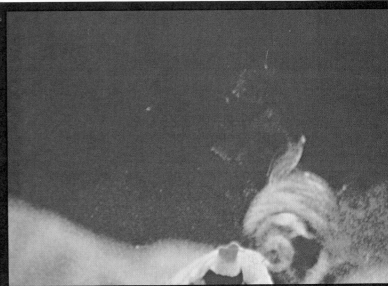

E/Victorious withdrawal. *The cloud of body fluids from the victim colors the water; the venom begins to affect the victim.*

F/To the victor the spoils. *In a few seconds the deadly venom has paralyzed the victim which will die within a few minutes. The snail will then eat.*

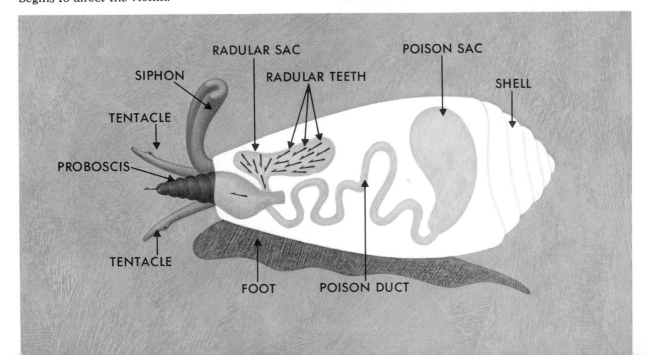

SIPHON

RADULAR SAC

RADULAR TEETH

POISON SAC

SHELL

TENTACLE

PROBOSCIS

TENTACLE

FOOT

POISON DUCT

Anemone—Passive Predator

Sea anemones, looking like flowers and growing attached to rocks and piles, are found in nearly all seas around the world, from shallow waters to depths of many thousands of feet. They prey on passing planktonic animals, worms, fish, crabs, and even sea stars. When a *Tealia* anemone feeds on tiny food particles, it expands its many delicate tentacles and pulls the food into its mouth. When a larger animal comes into contact with it, the anemone discharges the stinging nematocysts that line its tentacles. The nematocysts deliver a toxic substance, powerful enough to paralyze the animal so that it can then be eaten by the anemone. The clownfish is able to wallow in the tentacles of the *Stoichactis* anemone apparently unaffected by its stinging cells. The fish is able to secrete mucus which seems to inhibit the discharge of the anemone's stinging cells. Clownfish that have had their mucous secretions wiped off are seized by the anemone as they return to a previously harmless home. When isolated from its host, a fish requires a period of acclimation for the production of its protective mucus. The anemone may receive some benefit from the relationship since any predator that pursued the clownfish into the mass of stinging tentacles would be captured and become prey of the anemone. Clownfish may remove digestive wastes and parasites from the anemone.

Tentacles. Rows of short, tapering tentacles (above) surround the oral disc of the Tealia *anemone. The petallike appearance of the tentacles belies the fact that they are studded with stinging cells.*

The sting of anemone (right) is potent enough to kill most small animals that bump into it or that are presented to it, such as the bat starfish shown in the photo opposite.

Birds Feeding

Seabirds, flying high above the ocean's surface, can see for miles in every direction. They are alert for any disturbance indicating that small fish are being chased from below by larger predators. When attacked, the small fish are driven toward the ceiling of the sea, where they churn the surface waters in their desperate attempts to escape. The predators, in hot pursuit, may also break through the surface, increasing the turmoil. Shearwaters or other seabirds, seeing the splashing of the fish, fly over, knowing they may be able to find a meal. When the birds find a school of small fish trapped near the surface, they begin an orgy of feeding. Caught between two predators, the small fish are easy targets, and both predators enjoy good hunting.

From their vantage point high above the

surface, the hungry birds get an overview of the desperate battle. They can anticipate the escape routes of the fish. They can also judge when the fish are nearest the surface and can be most easily captured. The confusion soon reaches a fever pitch, with predatory fish lunging upward to their intended victims, and the birds diving at them from above. The prey frantically zigzags and jumps to avoid being eaten. Then, as suddenly as it began, the battle ends. The sated predators from the deep leave, and the prey,

*These **sooty shearwaters** are feeding on a school of fish frightened to the surface by larger fish. Huge flocks of these birds, numbering 100,000 or more, often take off in the early morning and remain at sea all day in search of food.*

no longer threatened from below, returns to safer depths, out of the reach of the hungry birds. The waters soon become calm. Such dramatic feasts in the open sea take place every morning at dawn and, to a lesser degree, late in the afternoon.

Chapter III. Color Me Invisible

Camouflage creates deceptive appearances. Animals deceive for two main purposes: avoiding predation and obtaining nourishment. Animals may use camouflage defensively; by blending in with their environ-

> "Some sea animals are quick-change artists, transforming themselves in the wink of an eye. Others require a little more time—but still get the job done."

ment, they create an illusion of invisibility. As an offensive weapon, camouflage enables an animal to approach its prey undetected. With some animals camouflaging is a full-time occupation, with others a part-time one.

Probably the most widely used form of camouflage among fish is countershading, or obliterative shading. It renders a fish almost indistinguishable to a potential predator or prey as the animal's color blends into the surrounding water and matches its reflection of light. The dorsal surface of these fish is dark so that when viewed from above it will match the deep-blue water below. Conversely, the white or silvery belly of most fish renders them practically invisible from below by matching the light sea surface above. Disruptive coloration is another form of camouflage which is especially common among fish of coral reefs. It's common among many other marine animals too. This form is comparable to the crazy-quilt patterns painted on warships and military vehicles. Another form of camouflage is cryptic coloration. It is common among marine animals, especially those that dwell on the sea floor. To match a background many animals have developed the ability to change color or intensity of color. Some are quick-change

artists, transforming themselves in the wink of an eye; others require a little more time but still get the job done.

Directive and deflective markings are another form of deception practiced by the animals of the sea. These markings usually draw attention to the least vulnerable part of the animal; or they may draw attention away from a particularly vulnerable area. Mimicry is another outstanding example of deception. An animal may seek to look like something else so it won't be recognized for what it really is. In some cases the animal mimics something inedible. In other species an animal which is not poisonous might mimic one which is. Yet in others, a predator resembles something harmless.

Many creatures that are brightly colored when in hand are virtually invisible in their normal habitat. Their reds or greens or blues have counterparts in their aquatic backgrounds, enabling them to blend with highly colored substrates. But there are some animals whose colors have not been so simply explained. It may be impossible for us as humans to completely understand all the colors and markings of marine life. We can only see them through our own eyes and not through the eyes of a fish. We have a background and mental makeup entirely different from that of a snail or fish.

Protection by mimicry. As a means of surviving, some animals have developed a strong resemblance to another animal or inanimate object. Here, an amphipod crustacean (upper animal) has mimicked the appearance of a snail (lower animal). By doing so and by remaining close to the snail, the amphipod is frequently overlooked by predators. Mimicry is found also among some species of fish and other animals. In some cases, the imitator duplicates a poisonous animal, which is not usually preyed upon because of its toxicity. In other cases, it may mimic an inedible inanimate object.

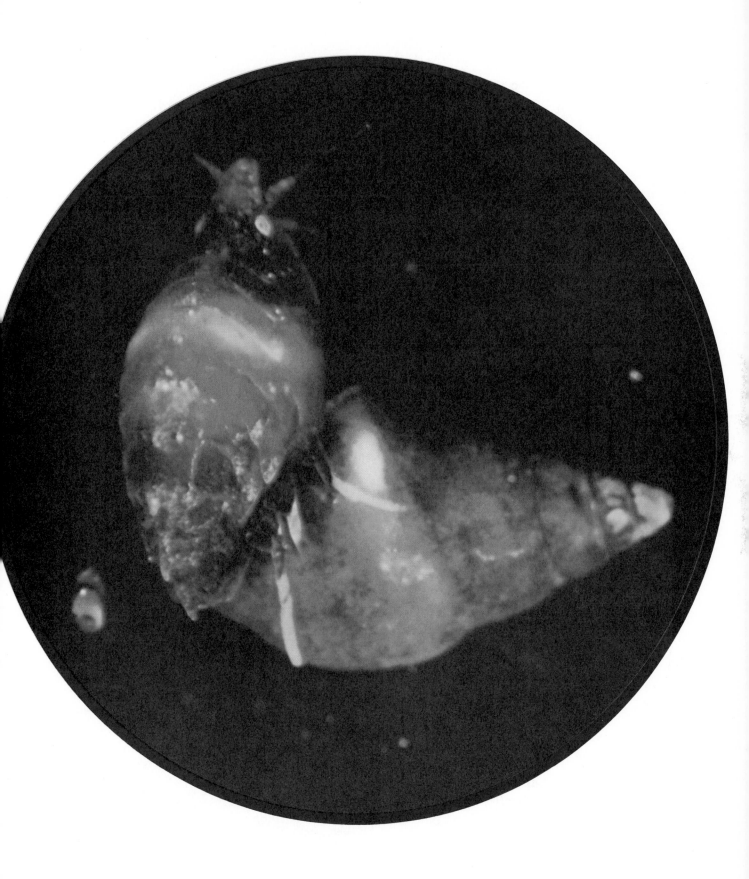

41

Fishing frog. Here we see cryptic coloration at its best. Deep in the Gulf of Aqaba, this fishing frog blends almost perfectly with a grayish sponge.

Israeli lizardfish. Lying in wait for its prey, the lizardfish depends on its cryptic coloration to avoid detection. It can strike with reptilian swiftness.

Find the Fish: Camouflage

The ability to change color helps some fish avoid detection. They change color to match their backgrounds, blending so completely that even a person knowing of their presence may have trouble finding them. This concealing kind of pigmentation is called cryptic coloration. It enables many sea creatures to see but remain unseen. Thus they can seize their unsuspecting prey as it passes by, or they can remain unnoticed by their predators.

It is remarkable, considering the variety of colors fish can assume, that there actually are only four variables within their color repertory. They possess only three pigments —black, white, and orange—and special re-

flective cells. It is these cells that permit fish to become an iridescent green or blue. Somehow they separate the spectral colors selectively reflecting some and absorbing others. Many fish have nervous control over the cells and can quickly alter the colors reflected. One of the most dramatic examples of this is the multicolored flashes that progress over the body of a spawning dolphinfish or mahimahi.

In the well-lit levels of the sea an adjustable color system is essential. Bottom texture and colors vary from one place to another as does the color of the surrounding water. But as one goes deeper the colors of fish are less important and as a result become more limited. The most logical reason is that deeper

Camouflaged sculpin. Blending with its pink background, this sculpin is barely visible even to a discerning eye.

Juvenile fish. Many young fish avoid being detected by predators, not through pigmentation, but by having no pigmentation and appearing transparent.

in the sea colors cannot be seen and therefore are of little use. This is because a particular colored pigment only reflects its own color of light. If orange light is not present in the first place, orange pigment will have nothing to reflect, appearing dark gray or black. Under these conditions a fish finding it advantageous to appear obscure or dark could be pigmented brown, black, or even red, as many deep-sea fish are, since red light does not penetrate deep.

In addition to cryptic coloration many bottom-dwelling fish possess bizarre body extensions and protrusions to assist in camouflage. The scorpionfish family is a master at this; some produce fleshy extensions on the fins and head with hairlike projections all

along the lateral line. The advantages derived are most likely a breaking up of the fish's outline and an irregular texture not unlike the algae-covered bottom on which it rests.

Another camouflage technique is that of body posture. The animal may assume a position which makes it less apt to stand out in contrast to its background. Pipefish are probably the best example. Their elongated shape, in a vertical position, closely resembles the marine grass in which they live. On the other hand, a horizontally swimming fish stands out dramatically against the background. Predatory trumpetfish often position themselves parallel to a slender gorgonian in an attempt to remain unnoticed.

43

More Hidden Fish

Corambe. At top, it is almost impossible to see the Corambe (right), a little nudibranch, nibbling on a lacy colony of bryozoans (left). The reason for the Corambe's resemblance to the bryozoans is hard to determine. The Corambe does not need to surprise the bryozoans, because these animals, like corals, live in rigid houses from which they cannot move. And since it has few enemies, the Corambe should not require camouflage for protection.

Klipfish. The mottled appearance of this clinid fish (immediately above) blends with its multihued habitat. Additional protection comes from the spiny dorsal fin rays that disguise its form.

Sea bass. At right, this mottled sea bass lies on a reef and is not noticed by the smaller tang swimming past. When the bass moves to another spot on the reef, it adapts by changing color.

Quick-Change Artist

The cephalopod cuttlefish, which is not a fish at all but a mollusc relative of the squid and octopus, is a master of disguise. It can change color in an instant or more gradually, whichever is more appropriate to circumstances. Unlike many other animals which can alter color only slowly through hormonal action, the cephalopods have nervous control over their skin pigment. Colored pigment is contained in cells called chromatophores and under stimulation they can expand or contract. What actually occurs is that many tiny muscles attached to the cells pull on the edges causing the cell to form a large flat plate making the pigment more apparent. Relaxation of the muscles causes the cell, and thus the pigment, to become concentrated into a dot and thus essentially invisible. The varied colors that these re-

markable animals can achieve result from a blending of pink, brown, blue, purple, and black pigments. Coloration reflects the mood of the animal—white for fear, red for anger, multicolor for sexual display.

Chromatophores also can exist in fish but differ in the mechanism of pigment dispersion. For example, the melanophores, dark pigment cells, have many branching extensions into which pigment can be forced. The pigment granules move while the cell remains basically the same shape. By a similar controlling mechanism, orange and white pigments can greatly increase a fish's potential for coloration.

Cuttlefish above has become an indistinct gray, enabling it to blend into its background.

Cuttlefish at right displays large, black spots prior to change, which will fade the spots into vague blotches.

Deceptive Eyes

The eye of a fish, even though it may be a target for attacking predators, cannot be altered much to avoid its conspicuousness. Tissues surrounding the eye and the iris itself may become pigmented to match the color of the body but there will always remain a jet black pupil. Because this black spot on the head is inevitable in all fish, many have evolved with color patterns that obscure the eye or detract from its prominence. For instance, a black bar on the fish that extends through the eye region makes the eye very difficult to distinguish and may be advantageous to its owner. Another even more creative color pattern is the eye spot near the posterior part of the body seen on many butterflyfish, flatfish, wrasses, damselfish, gobies, blennies, and even on some rays. One example of the effectiveness of such an eye spot is in the misguided attacks of an Indo-Pacific blenny that preys on other larger fish by tearing off pieces of skin. It often goes for the eyes, but when confronting a butterflyfish, it is directed to the eyespot at the rear of the body, which does much less harm to its victim.

Zanzibar butterflyfish. A distinctive directive mark may call attention of predators to less vulnerable parts of this fish (above). Black lines break up the animal's outline and give it added protection.

Golden long-nosed butterflyfish. Not only does this fish (right) have a false black eye at the rear end of its dorsal fin, but it also has stripes that help break up its outline and further confuse its enemies.

Two-spot octopus. Defining the function of spots which to humans resemble eyes, on this octopus (upper right, page 49) or any other marine animal, would be difficult. We cannot judge what another animal may see and how that observation affects its behavior.

Four-eye butterflyfish. The apparent false eye near this fish's tail (lower right, page 49) may draw the attention and perhaps the attack of predators away from the most vulnerable part of the fish—the head.

48

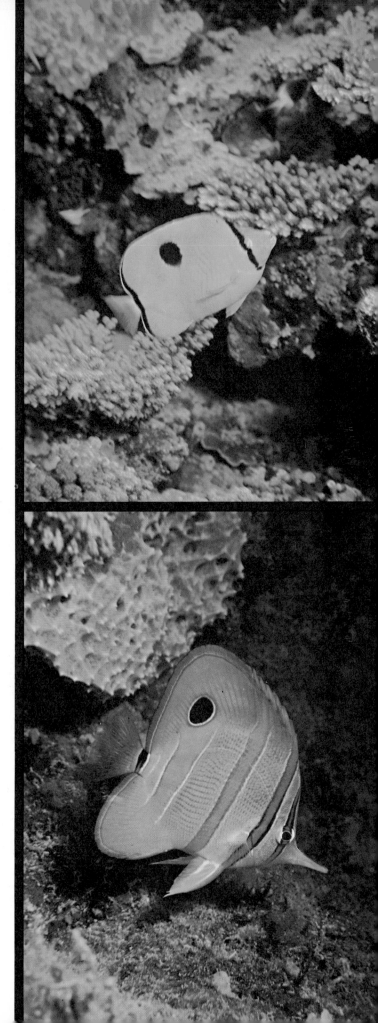

Moorish idol. Two nearly vertical black stripes deceive the eye and thereby make the actual form of this fish unrecognizable and difficult for a predator to attack at the proper angle.

Disruptive Coloration

Stripes, bars, and patches on a fish break up its outline so its fish form isn't obvious to an observer. To escape notice by a predator, a potential prey must divert attention away from itself as a whole. Successful coloration and markings of the prey present an optical illusion to the predator. A striking example of this is the reef fish which has black bands of color that extend along the sides of its body. When a predator strikes, the attack is usually directed at the front of the fish with the head as a point of focus. By obliterating that distinct feature with a black band, the prey is given some degree of protection. The

Butterflyfish. *This fish is a good example of disruptive coloration. The black on its head blends into the dark background of the reef it swims in. The yellow blotches on its body break up the white areas.*

role of light and distance on a fish's appearance is another factor to consider in terms of coloration. To a diver looking at reef fish with artificial light and at close range, the imposing light and dark markings are impossible to overlook. But under the naturally dim play of light, objects a few yards away are often indistinguishable, and that same fish appears to vanish as its colors blend together in the distance. Another aspect to consider is the image seen by a predator. Studies have shown that many fish do not see as well as man and that some cannot even see different colors. With such handicaps a predator must be an amazingly efficient organism to be successful.

Stripes That Hide

To a schooling fish any markings that confuse a predator will increase its chance of survival. One effective means of confusion is having dark and light stripes which obliterate the distinct outline of each individual fish. The stripes of schooling fish are most often horizontal, but a number of vertically striped species do exist. In a school of striped fish an attacking predator is faced with the problem of identifying an individual prey

Vertical stripes. In open-ocean fishes such as these jacks, bold bars on their sides may hide them. Although such stark markings may be highly visible close up, they fade into gray when viewed at a distance in the open sea, especially when the fish are schooling. The light and dark areas blend together when seen through the water and make the school a hazy mass to the observer.

and then pursuing that fish until its ultimate capture. But all that is visible is a writhing mass of flashing lines; there is motion everywhere, and no single fish can be differentiated from its neighbor. Hundreds of eyes and hundreds of lateral lines perceive the predators at every turn, all fish responding to it individually, none colliding. Some people have described a school of fish as a superorganism in which many individual units, all the same, function together for the good of the whole. In a sense this analogy is valid because many schooling fish could not survive alone if they were separated from the group. A school appears to be one large fluid unit, rather than hundreds of small, vulnerable animals.

In a school there are more individuals to sense a predator, which may be some advantage to the group. The compact schools of some small fish may actually deter predators by appearing to be an object too large to attack. Some statistical research has shown that a school of fish in a given area will have fewer chance encounters with predators than if the individual fish were evenly dispersed. This means that a predator, assuming he could eat only so much at a time, would catch more fish over a period of time if they remained separate

There are a number of other advantages fish derive from living together in schools. More eyes and nostrils are available to seek out or detect food. Some fish change color as they begin to feed, possibly as a stimulus to others in the group to begin feeding. Many schooling fish feed on plankton, and their prey is far less likely to escape from a school than from a single fish. If a copepod darts away from one fish in a school, it will almost certainly find itself in the path of another. One further advantage for schooling fish is that no complex system of strenuous migration is necessary to bring the sexes together for reproduction. Breeding can take place merely by the simultaneous activities of the males and females. Inconveniences to schooling are rare but obvious—for example, when barracudas herd a school of sprats, feeding upon them for weeks until extinction.

*Vertically striped **damselfish,** such as those above, are highly territorial, and at the slightest hint of danger will race to hide in the coral reefs they usually inhabit. As we can see in the bottom of the photograph, some of the fish have already chosen to conceal themselves in the coral. The large fish in the photo's center may be the reason the little damselfish are trying to hide.*

Grunts and jacks. *In the photograph covering the next two pages, the grunts, below, with their horizontal stripes and vivid coloration, remain close to the sun-dappled bottom where they can more easily blend in with the colors of the undersea terrain. The jacks, above them in midwater, are typically countershaded—darker above than below.*

Blending with Background Light

The silvery color of these anchovies, which are being chased by a yellowtail snapper, results from a reflecting layer in their scales. The layer is composed of iridocytes, opaque crystals of a waste product called guanine. These crystals reflect light in different ways, providing a silvery color at times and a white appearance at others. Additional layers of iridocytes and a mixture of iridocytes in the layers of normal pigment result in irides-

cence. The iridescent quality of butterfly wings probably results from a similar organization of pigment and reflecting material. How the light is actually reflected is not certain, but it may be that closely packed parallel layers of material allow certain wavelengths or colors to be reflected at a particular angle while others are absorbed. In any case, when light reflecting off fish which have iridescent coloring is close to the

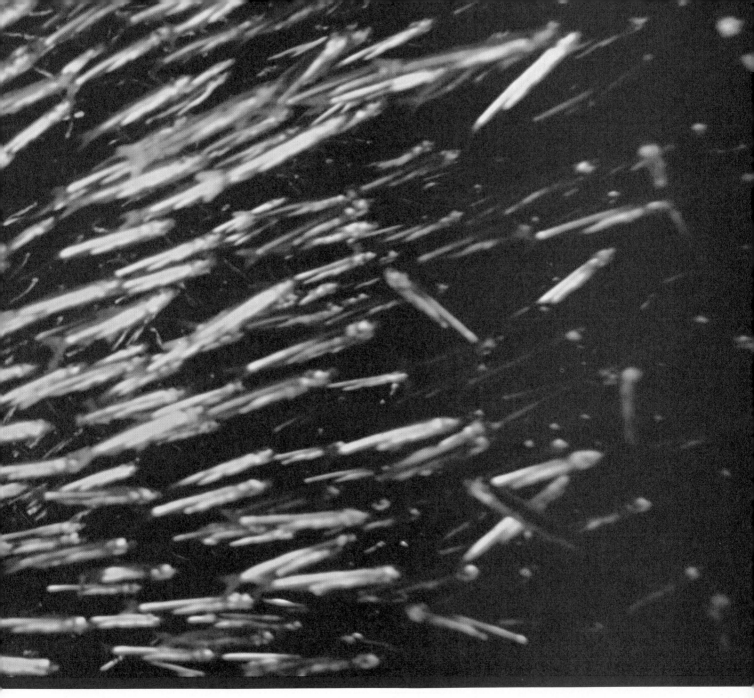

intensity of the natural background light, they become nearly invisible. But this yellowtail is close enough to see these anchovies. If it misses getting one, it will probably be because it couldn't zero in on a single fish since so many are present. The play of light from a group of these fish is a beautiful spectacle, but trying to keep an eye on one individual in such a shimmering mass is almost impossible. This is the intended effect and is a distinct *advantage* in confusing an attacking predator. Experiments with

Yellowtail snapper and anchovies. These anchovies, chased by a yellowtail, appear silver due to their reflecting layer of scales.

schooling fish in which a few were abnormally colored with a dye have shown that predators immediately select them as targets and are more successful at capturing them than normal fish. In other words, predators weed out any abnormal fish, thus making the species as a whole better able to survive. The "fittest," in this case, are those that blend with the school.

Cooperative Roundup

To a fisherman peering into the water almost all fish are invisible either due to the opacity of the water itself or the countershading of the fish. Because of this, his success may depend upon his ability to observe signs indicating the presence of fish. A school of fish feeding or being fed upon often disturbs the surface. Birds circle above and dive periodically amidst the animals scrambling for a meal. Their activity can be seen for consider-

able distances and act as a sign to fishermen. For instance, in this Mauritanian beach scene, man, bird, and beast combine efforts to garner a harvest of mullets. These schooling mullets may have been chased inshore by other animals or may have been feeding there. Perhaps the seabirds above saw them first and inadvertently signaled the fishermen below. But when the men beat the water with long sticks, dolphins drive the school of mullets into the nets of the fishermen and simultaneously feed on them. The

seabirds pick off pieces of fish left over by the dolphins. Soon the school is decimated.

There are other indirect methods of locating fish schools that to some extent compensate for the invisibility of fish to fishermen. Areas of dense plankton populations can be easily sighted by the color of the water and may indicate the presence of useful fish. At night the bioluminescence caused by schools as they swim through the water is another, if more limited, indication. In addition to these

Working together. In the photograph above, fishermen are rounding up a school of mullets. Sometimes circling birds give the fishermen a clue as to where to find a school of fish, and sometimes dolphins assist by herding the fish toward shore.

natural signals, man employs a host of scientific equipment to pinpoint fish far below the surface. Fishermen must depend on schooling behavior for large harvests of fish. Consequently the normally advantageous schooling way of life becomes a liability in terms of man's impact on the sea.

Chapter IV. Living in Armor

In the undersea world of predator and prey one of the best defenses is armor. And animals display a wide variety of armor, some of which man has been able to copy very effectively in his own weaponry.

But animals that adapt to life in armor face a major problem—they are cumbersome and slow-moving.

Animal armor comes in several styles. There are exoskeletons—skeletons that are external instead of internal as is man's. There are tubes, shells, and cases that are part of the animal. Exoskeletons are widespread even in the plant world. Diatoms, basic ingredients in the ocean's intricate food webs, have silicate shells which enclose the cellular components of these plants. Other microscopic marine plants—the silicoflagellates—have shell-like plates protecting them. Many seaweeds secrete and deposit on their exteriors a coating of lime that armors them.

In the animal world, the examples of exoskeletons are legion. Some of the one-celled protozoans, sponges, and corals—including the stinging corals, which are hydrozoans, and reef-building corals, which are madreporarians—have exoskeletons. The armor of crustaceans, including lobsters, crabs, shrimps, and barnacles, is familiar to many. And molluscs, like clams, scallops, mussels, and oysters, are also well protected.

Then there are the creatures that are born without armor but eventually live within cases, tubes, and tests of their own making. Some armors start out as soft mucous secretions; these secretions combine with lime solutions and develop into tough outer coatings, which can ward off physical or chemical attack. Some other creatures that are born without armor inhabit the abandoned shells of other animals. Or these creatures make a safe haven in the substrates. They find protection as they gradually envelop themselves in the substrate.

In a class by themselves are certain fish and reptiles that are clad in a different kind of armor—scales, which can be thick and tough. Some fish, however, have evolved tough, scaleless skins.

Man's most obvious imitations of these animal armors are our military tanks. But there

> "Man's most obvious imitation of animal armor is the military tank. But there are other examples: ancient ironclads bristling with steel stakes, contemporary battleships, armored cars, air-raid shelters. And all buildings are a form of armor."

are other examples, like early ironclad naval vessels, which were armor-plated and often bristled with steel stakes. Contemporary warships have plates of the toughest steel. And armored cars, which travel about our city streets with bank funds, are covered with almost impervious plates. Air-raid shelters in warring countries are a form of armor. And all buildings in a sense are armor.

Dual armor. Protective armor comes in several forms. This pencil urchin, an echinoderm related to the sea star, carries stout clublike spines on its rugged calcareous outer shell (or test). Its spines are tough enough to ward off the attacks of many animals. And to repel the predators that manage to get through this part of its armor, the pencil urchin has a second line of defense—an inner test, which encloses its vital organs. Yet a few animals, like crabs, sea stars, and large fish, can penetrate the pencil urchin's dual armor and feed on it.

Heavy Armor

Molluscs owe much of their success to the heavy armor they carry about. Over 500 million years ago their ancestors probably possessed a simple horny covering for protection. They subsequently gained the ability to impregnate this covering with calcium carbonate. This shell, along with a strong muscular foot, may have allowed snails, the largest and most successful group of molluscs, to exploit habitats too inhospitable for other animals. Fossil evidence indicates that much of molluscan evolution took place in the shore zone where an abundance of food and a variety of habitats existed. An impervious shell would have provided protection from both drying at low tide and abrasion, and a strong muscular foot would have enabled them to hang on tightly to wave-swept rocks. One mollusc that successfully endures such a habitat is the

limpet, a small animal which possesses a pyramid-shaped shell. The pointed shell probably reduces the unsettling effects of waves, allowing the animal to graze on algae in the intertidal zone where their greatest force is exerted. Some of the primitive gastropods had coiled shells but originally they were coiled in one direction or plane, much like the shell of a chambered nautilus. Evolutionary modifications of this shell have made it more compact and better balanced and have resulted in the spiral conical configuration seen on many snails.

In addition to the protective coiled shell seen on most gastropods, a horny "trapdoor" or operculum further isolates the animal from outside. This durable shield may be any shape depending on the aperture of the shell. The queen conch has a curved narrow operculum that fits way back inside the shell while the turbo snail has a beautiful, circular calcareous operculum often called the "cat's-eye."

A hard shell is an effective protection, but these molluscs, however, are not invulnerable. A number of predators are successful at circumventing this deterrent. Bat rays possess platelike grinding teeth able to crunch the heaviest shelled clams; some starfish extrude their stomachs to digest the protected snail; others ingest the whole snail.

Eggshell cowry. With its pure white, glossy exterior this cowry (opposite page, left) stands out clearly against almost any background.

The same cowry (opposite page, right) has covered its white shell with its black mantle. The mantle can be extruded to protect the shell from other organisms settling and attaching on it.

The triton on the right is large, heavy, and somewhat disguised by the growth of algae on its shell.

The same triton, below, withdrawn into its shell, displays his horny operculum (trapdoor), an additional protection from attackers.

Armor

The Crustacea are a group of animals with a hard armorlike skeleton on the outside of their bodies. Physical protection is the most obvious advantage to having an exoskeleton because it permits some crustaceans to lumber about in broad daylight, caring little for danger from others nearby. One disadvantage is the inability of a rigid outer covering to accommodate growth. As a result the vital protection must be cast off periodically and a new larger one constructed. During this time of transition the animal is extremely vulnerable and must remain hidden until the new shell hardens. An exoskeleton also

Pacific rock crab. This animal has an exoskeleton armor, as do most other crabs. Because of this hard exterior, the rock crab must shed its exoskeleton periodically, thus enabling it to accommodate its normal body growth.

limits maneuverability for many animals. But others with a lighter covering (such as the mantis shrimp) are surprisingly quick and graceful in their movements.

Some crabs find it advantageous to inhabit the empty shell of a snail rather than fabricate a complete covering for themselves.

These aptly named hermit crabs have a soft but muscular posterior that is curled to fit the spiral of their borrowed homes. They too must seek new protection as they grow, and when moving to a new larger shell, they must expose their unprotected body to predation. The muscled abdomen makes them surprisingly difficult to pull from a shell.

Hermit crab. This crustacean has a soft body. To protect itself, it adopts the abandoned shells of other sea creatures, and as it grows, it searches for larger abandoned shells. It can be found in shallow waters around the world.

Miniature Monster

The horseshoe crab pictured above is the epitome of an armored animal with its chitinous, rounded shell adorned with barnacles on its back and with a long, spikelike tail. This animal, although commonly called a crab, is not really a crustacean but is more closely related to the arachnid group, making it allied to spiders rather than crabs. Unlike the crabs, this primitive arthropod possesses a tail, which it uses in righting itself when it is overturned. It is also a much slower and more awkward swimmer. These primitive creatures, whose form has remained unchanged for some 200 million years, may burrow into sand for extra protection. The armored adults contain very little meat and are surely unpalatable to most predators, but the young do succumb to some birds and fish. As the young grow older and larger, they molt periodically. When they are full grown they keep their armored shell, as seen above.

Loggerhead Turtle

The body of this loggerhead turtle, like that of other sea turtles, is protected by the bony armor of its shell. The shell is made of two layers. The outer layer of plates is derived from reptilian scales and the inner layer has its origin in bony tissue. The points of junction for the first layer do not lie directly above those of the bony layer, and this gives the shell extra strength. Unlike the exoskeleton of crustaceans, the turtle shell grows at the edges of individual plates of both layers, and this growth can be measured by rings or markings on the shell. The turtle continues to grow throughout its life, and its shell grows along with it. The heavy scales of the skin covering its flippers and head, plus the hard bone of its skull, afford additional protection. When attacking, the toothless sea turtles use their sharp-edged jawbones, which are powered by strong muscles. Their broad flippers enable them to move well in the sea where they have few enemies.

Pincushion and Box

For animals that are not built for speed, other protections have developed. Two families of fish have solved the problem in a similar way: their bodies are too hard for a predator to handle.

The spiny puffer's body is densely covered with short, sharp spines, which are actually modified scales. When it is threatened, the puffer can inflate itself with water to make itself more imposing and to erect its piercing spines. The puffer has another defense: it has poisonous flesh. The poison is a neurotoxin, affecting the victim's nervous system. In man it shuts down signals from the brain to the diaphragm, so breathing becomes impossible. Marine animals may also die after consuming puffer.

The boxfish and its relatives, the cowfish and the trunkfish, have developed a hard shell-like outer covering that approximates an exoskeleton. Besides their outer armor, boxfishes and their relatives rely on a poison they secrete as an additional protection. The bright color of the boxfish may serve as a warning to predators to stay away from this crusty little fish.

Threatened puffer. In the face of danger, the spiny pufferfish, above, inflates itself with water and erects its sharp spines. Its pincushion appearance is an effective deterrent to most of the pufferfish's enemies. Unfortunately, it backfires when the predator is man, who finds the inflated bodies of pufferfish and their relatives attractive as lampshades or curiosities.

Immobile boxfish. The hard outer covering of the boxfish protects it against attack from larger predators. However tiny parasites are sometimes able to get under its skin. At the right, a little neon goby seems to be nipping the immobile boxfish, but may be picking a parasite from it. Boxfish are usually found in tropical waters, where they feed on some of the smaller invertebrates.

Spines That Protect

Some fish have sharp protective spines. These spines may occur almost anywhere on the body and in some cases occur everywhere on the body. Spines are either modified scales or spiny rays of the fins or bones that project out from the fish's body. A classic example of such modified scales is found all over the spiny puffer. Surgeonfish and tangs also possess scales modified as razor-sharp scalpels which evolved from bony ridge scales on the caudal peduncle. These lancelets are attached at the posterior end, projecting forward, and can be erected or depressed at will. Another example is the dagger on the tail of the stingray.

Spiny rays of the fins have also a dermal origin and possibly developed from some kind of scale. Supporting the idea of an evolutionary development toward spiny rays is the fact that almost none of the primitive fish possesses stiff rays while many of the more advanced fish do. These spines are of great importance for the defense of many fish, acting as instruments for the injection of poison, making the owner difficult to swallow, or merely providing an unpleasant stingy surface to deter enemies. The development of elaborate spinous fins are a characteristic of the scorpionfish family, and spines are notorious in the lionfish and turkeyfish and in many other bottom-dwelling fish. The dorsal spines of the weeverfish are actually

used in offense—it has been said to attack divers with its venomous weapons. Spiny rays can be found on the dorsal fin, the pectorals, the pelvics, and the ventral fin—or on all of these. In some cases the mere erection of these spines as a threat may deter a predator. Some reef fishes, like the triggerfish or the filefish, are able to erect their spines in such a way as to make them impossible to pull from the hole in which they are hiding.

Among the fishes with bony spinous projections the scorpionfish are probably the best known. Covering their heads are sharp bones that project outward, rendering them safe from most predators. The spines may help these bottom-dwelling, well-camouflaged creatures in breaking up the outline of the fish and making it look like another irregularity on the background. Probably the most common part of the fish's body where defensive bony projections have developed is the operculum or gill cover.

French angelfish. These fish (left) have spines on their gill covers which they can fan out to fend off attackers. Other angelfish also have these cheek spines.

The Achilles tang (right) has a scalpel-sharp spine on both sides of its caudal peduncle, the portion of the body just in front of the tail. The spines are at the back end of a brightly colored heart-shaped patch. This flash of color probably serves as a warning to attackers.

Hard to Swallow

The shapes of some animals make them difficult to swallow. Certain deep-bodied reef fish, for example, have flattened body forms which makes them impossible to consume in one mouthful. A predator such as the grouper, because it lacks the ripping and tearing teeth necessary for biting off portions of its prey, must be able to engulf its victim whole and then "chew" it up internally with its pharangeal teeth. This feeding method excludes any fish which cannot be taken in one mouthful.

As a result, deep-bodied fish such as the angelfish, butterflyfish, surgeonfish, tangs, filefish, and triggerfish may receive a certain amount of protection from predators like the grouper. This body shape has other defensive advantages. A short deep-bodied fish can maneuver very easily among coral heads on a reef, possibly fooling predators. This body shape, coupled with their vivid coloration—surgeonfish range from yellow to purple, filefish are sometimes a deep brown with white and black spots—makes it easy for these creatures to conceal themselves among the multicolored coral.

Seahorse. The irregular shape of the seahorse (left) gives predators a difficult task. Since most animals take in their prey headfirst, an attempt to ingest a seahorse must include maneuvering it into the right position. This often proves to be a hard job. The seahorse also has a tough and leathery, though scaleless, skin to protect it.

Trumpetfish. With its elongated shape and bony head, the trumpetfish (top right) is not likely to be taken whole by another fish. When a predator tries to swallow it headfirst, the trumpetfish resists and usually must be bitten into chunks to be ingested.

The long-nosed butterflyfish (lower right) has a shape that a predator can't swallow easily. But at least two other factors also help it survive. Its disruptive coloration disguises its fish shape and makes it less readily recognizable as fair game. And its dorsal spines, erect when the fish is threatened, can stick in a predator's throat.

Chapter V. Strategic Withdrawals

We are all familiar with Goldsmith's lines: "For he who fights and runs away/May live to fight another day." This statement has truth for life in the sea as well as on land. An

> "For he who fights and runs away
> May live to fight another day."

animal outmatched in a fight is wise to withdraw if it can. Some creatures spend their lives hiding on the sea bottom or living inside a shell, so their very existence is a form of strategic withdrawal. Others, less sedentary, are candidates for attack from all sides, and when threatened, many prefer to leave the battleground. But how do animals escape especially if a predator continues to pursue them? Most simply, some escape by turning and fleeing, outdistancing or outmaneuvering their opponent. When we think of one animal outrunning another, we usually think they must have great speed and so it sometimes is. But many animals we consider to be incredibly slow-moving can move just fast enough to outrun an animal seeking to make a meal out of them. On land, a fugitive tries to escape in two dimensions. In the sea, an unpredictable three-dimensional sharp turn gives the advantage to the pursued over the pursuer, even if he is substantially slower.

> "Many animals we consider to
> be incredibly slow-moving
> can move fast enough to outrun
> an animal seeking to make
> a meal of them."

Some burst out of water in flight, returning every few seconds to scull with their tails.

Then they are off again. If the deck of a ship is not too far above the surface, a variety of sea animals is sometimes found there in the morning. Surprisingly, some creatures not even known to be "flyers" show up, thus giving us new insight to their ways and habits. Until the voyage of the *Kon Tiki,* authorities generally ignored reports that squids jetted themselves right out of the ocean. But with the evidence gained on this voyage, they began taking a closer look at these remarkable molluscs.

Before the development of radar, man-made vessels could also hide. If a vessel was losing a naval battle and a fogbank was nearby, it might make a run for the fog hoping to escape its enemy and defeat. Some animals like octopus, squid, or sea hares use a similar tactic when they are confronted by a superior foe. While they do not have fog to hide behind or in, they release an ink cloud that is equally effective. And since they have control over its liberation, they can decide when it should be brought into play.

When threatened, some marine creatures which live in the substrate or among plants may duck into pockmarks in coral reefs, cavities in rocks, or other holes. Or they may bury themselves in the sand.

The animal that knows when to flee and when to stand and fight is usually successful in its continuing struggle for survival.

An octopus flees before the benign pursuit of a diver from the Calypso. *Some maneuvers the octopus may use to escape danger are changing color to camouflage itself, retreating into a hole its pursuer can't get into, or releasing a cloud of dark ink to confuse and mislead a predator. Or the octopus may simply jet away as it is doing here.*

Jetting. *A sea star (lower right) approaches two scallops. One scallop escapes the predatory sea star by bursting from the bottom in an explosion of sand.*

Jet Propulsion

Encumbered by its shell, a scallop lives a fairly stationary life, spending most of its time on the bottom. But the shell is not an impregnable fortress, and when the scallop detects the approach of a predator, usually a sea star or an octopus, it flees, using its jet propulsion mechanism. Surprisingly enough, the direction of movement is forward with the open valves of the shells facing ahead. This is because as the shells clap shut, water is forced out of small spaces on either side of the hinge. The jetting scallop's movement is jerky; and the scallop cannot escape a hungry octopus, but it can outdistance a sea star.

*Wild gyrations carry this pastel **nudibranch** away from the reaching arms of a sea star. Survival through escape in a three-dimensional world.*

Gyrations

The nudibranch glides easily over the substrate, going about its daily routine. A predatory sea star, attracted by a chemical given off by this delicate animal, moves in for the kill. The sea star is a slow-moving predator, but its speed compares well with that of the nudibranch. Silently inching toward its intended victim, the sea star reaches out with one arm and touches the nudibranch. When contact is made, the nudibranch reacts. It gyrates wildly, trying to get off the sea floor and out of the grasp of its predator. When it finally begins to rise, it continues its gyrations until currents carry it to safety.

Flyingfish. This fish has a remarkable way of escaping predators. With a burst of energy it flies out of the water and glides above the waves for as long as twelve seconds.

Escape by Flight

In warm waters and far from land, this flyingfish, a juvenile in the Sargasso Sea, bursts through the surface of the sea when chased from below by predators. The attempt of flyingfish to escape to a safer locale brings them briefly into our world. At the approach of danger they accelerate toward the surface. When they gain enough speed to get airborne, they leave their predator behind. The fish skims over the wave tops, gliding first in one direction and then another. It may look down before landing; if it sees a predator still following, it sculls its tail in the water and scoots off in another direction.

*These two **mudskippers**, one partially hidden behind the plant, cling to a rock. If a predator attacks, they attempt escape by outrunning it or by jumping out of the water.*

The mudskipper is an unusual fish, since it is almost as much at home out of the water as it is in. It feeds on the mud flats exposed by receding tides. When it is alarmed, it bounds with great agility across the flat, skipping in a series of short, quick leaps. There it "walks" on modified pectoral and ventral fins, hopping about until the danger is passed. When it finds a crab's burrow or other crevice to hide in, it stays there until it feels safe.

When the mudskipper is threatened in deep water, it leaps out of the water. But it is not as skilled at staying airborne as the flying-fish, and it plops back only a short distance from where it exited.

Jumping to Safety

Adult barracuda tend to be solitary hunters, while younger ones hunt in schools. These sleek, streamlined predators are machines of death for the smaller fish on which they prey. They have been observed herding fish into a compact group and then cutting a swath through them, snapping at and killing their prey without taking time to eat them. Then they turn around and eat at their leisure. Often the barracuda's attack is so swift that the herded fish cannot react to the marauders and become easy victims. But when the prey senses the coming attack, the schooled fish may leap out of the water in flight. In the air they are momentarily safe from the snapping, slashing teeth of the barracuda, but when they plop back in, they are again in danger. Flight reactions are not uncommon when schools of fish are attacked from below. At times it is possible to tell the path of a predator by the sequential jumping of fish in a line across the surface. Some of the

fish which become airborne most often are the halfbeaks, the needlefish, and, of course, the flyingfish. These are all sleek, surface-dwelling fish which seek the nearest protective exit when threatened. Another marine animal prone to jumping is the squid. Squids have been reported to leave the water at dusk with amazing speed and hurtle through the air for a considerable distance.

There are other jumping animals which use their ability to jump for reasons other than

*The **barracuda,** sometimes called the wolf of the sea, may hunt in packs and herd its prey into a compact group.*

safety. The manta rays are marvelous jumpers, sometimes becoming so active that five or six may be in the air at the same time. Whales are another group that thrust their great hulks up out of the water then fall crashing back. Some think that such behavior is associated with a kind of courting ritual, but these impressive feats do not always occur during the mating season.

Protection by Boring

Some animals spend nearly all their lives in hiding. Boring molluscs find safety by tunneling through mud, wood, and even rock.

Most clams burrow into sand or mud using only their soft foot as a digging implement. In Puget Sound along the coast of Washington one species of large clams, the geoduck, may weigh twelve pounds and has a body long enough for its neck to reach the surface from a burrow as deep as four feet.

The largest and most efficient rock borers are the pholad clams. Near the end of their larval stage, the clams fasten themselves to a surface of heavy clay, sandstone, or limestone. They begin to make their burrow when their shells begin to form and harden. They move from side to side, or up and down, rasping the burrow face with the shells until a hole is gouged out. The movement and grinding continue, and soon the hole becomes a tunnel, which the clam increases in length and diameter to accommodate its own growth. In fact the mollusc be-

comes completely trapped within its chamber. The orientation of the shell is such that it can burrow forward but cannot reverse the direction. Thus the protective chamber becomes a prison with its only connection to the outside world being two siphons to the opening of the tunnel—one to bring in water and food and the other to expel water, body wastes, and gametes. Many animals become so precisely adapted to their specific confined environment that they are unable to survive anywhere else. These burrowing clams are no exception, and possess reduced shells

Pholad clams. *The shells of boring molluscs do not begin to form until the larvae settle on a surface they can bore into. In maturity the spherical shells are an efficient grinding apparatus, and the molluscs can scrape rock at the rate of an inch per year.*

which are useless for protection because they are not large enough to cover the body. If released from their chambers they would be extremely vulnerable and unable to survive in the open. Boring molluscs are well-concealed within their tunnels. We often don't know of the presence of wood borers until the pier collapses.

Feather-duster worms and Tealia *anemone.*
On the left are three open feather-dusters in feeding
position. On the right is an anemone.

Closing Up

The feather-duster worm and the *Tealia* a-
nemone are sessile animals that live fastened
to the substrate. When they are threatened
or disturbed, they cannot move away easily,
so they have devised other systems of "escape." The worm has very fragile gills, which

it extends for feeding and respiration. If they
are damaged, the worm might die. Fortunately the feather-duster secretes a rigid,
fairly strong, tubular structure around itself.
So, when the worm senses danger, it quickly
withdraws into its tube, where it remains
until the danger is passed.

The anemone does not construct a tube or

Withdrawal. *Having sensed danger, the feather-dusters have withdrawn into their tubes. The anemone has folded its tentacles and is fully closed.*

other protective device, but it can withdraw into its own body cavity. When danger threatens, the anemone folds its tentacles toward its oral disc and then rolls them inside, until the sensitive tentacles are covered. After a short while, the anemone tentatively begins to open up again, fully extending itself if the danger has passed.

"When the worm senses danger, it quickly withdraws into its tube, where it remains until the danger is passed. It has very fragile gills, and if they are damaged the animal may die."

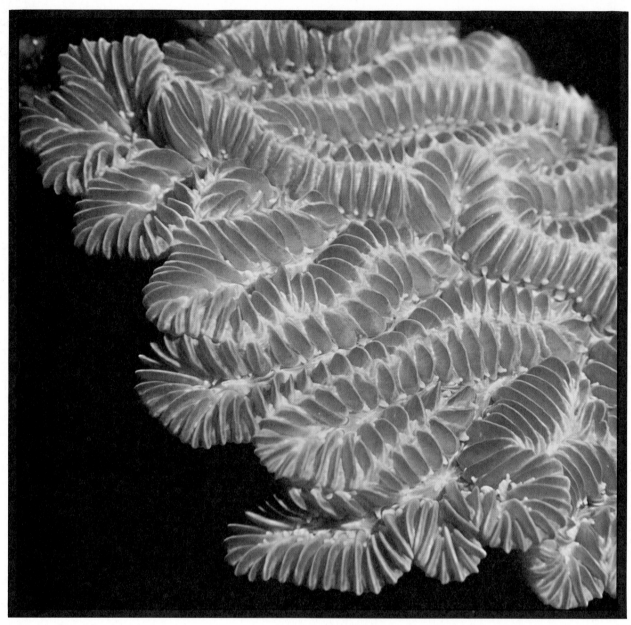

Living in a Castle

Few divers exploring a coral garden are able to observe the full beauty of these delicate animals. Unless the flowerlike corals are extended to feed, which generally happens at night, the corals hide inside the limestone castle they've built for themselves. Even more striking than the difference between the open and closed polyps is the dead, cleaned skeleton of a coral compared to the living specimen.

In a living coral the animal completely covers the hard calcium carbonate skeleton giving little indication of what intricacies lie below. A dead specimen whose tissues have been cleaned off shows the delicate skeletal

"Coral polyps are reasonably safe behind their hard walls. But a few predators can penetrate these defenses and feed on the flesh inside."

septa or projections that were secreted by the animal itself. These individual imprints are the easiest way to differentiate between species. However hard their protective skeletons, a number of reef fish, among which are triggerfish and parrotfish, have teeth or beaks hard and powerful enough to bite off chunks of coral, grind it in their throat, digest algae living on the coral, and eliminate fine sand.

When the little animals stretch out to feed, as most do at night, they use their tentacles,

Hidden. *The tiny animals of this small section of a brain coral colony (left) have retreated to the safety of their limestone castle.*

Out to eat. *Delicately colored brain coral polyps (above) wave stinging tentacles to trap food. They retreat at the slightest disturbance.*

which are lined with stinging nematocysts, to entrap tiny planktonic animals. The poison of most corals is effective only against the minute forms that make up their diet and not against larger predators.

Tridacna clams. *The shells of these three tridacnas are hidden amid a garden of sponges. In this picture their colorful mantles are exposed.*

Sealed Tight

The *Tridacna* is a fabled resident of the coral reefs of the Pacific. Legend holds that this large mollusc, sometimes four or five feet wide, is a killer clam, grabbing careless pearl divers between its two huge shells and holding them until they drown.

But this is all fantasy. The *Tridacna* feeds largely on algae, which grow in its mantle between its shells. The shells are left agape to allow sunlight to reach the plants so that photosynthesis can occur. Tridacna clams have such large mantles, expanded for more exposure to sunlight, that they cannot close their shells fast enough to surprise a diver.

Bag of tricks. A sea cucumber has some very effective defenses against attackers. Here it ejects skeins of sticky white mucus to immobilize a starfish.

Multiple Defenses

As they inch their way along the seabed, sea cucumbers may seem extremely vulnerable. In truth their defenses are formidable. Predators are discouraged by the poisonous skin of some sea cucumbers. A cucumber that is disturbed reacts by expelling water from its body and contracting itself. A truly desperate sea cucumber resorts to a remarkable defense: it turns itself inside out, spewing out its insides—respiratory and reproductive organs and even its intestines—which entangle its hapless attacker while the cucumber escapes. In about six weeks the eviscerated organs are regenerated.

Hiding in a Coral Head

Tiny, bright-blue damselfish live in sunny tropical waters. Several hundred of them may inhabit a coral head, swimming just above it during the day and picking bits of food from the water. Generally the distance they swim from the coral head is indicative of the degree of danger around. The distant approach of a diver may cause them to move closer to their protective coral but not until the diver moves to within a certain critical distance will they instantly withdraw and hide in unison. It is remarkable to see how many of these apparently defenseless fish can completely disappear into a coral head out of reach of almost any predator. Because of their life-style some species of damselfish never grow to any considerable size, therefore they can continue to live in their coral habitat. Individuals growing too large would

be more easily captured by predators. Even if the coral head is removed from the water, the little fish does not leave it, but relies on its security to the last.

The arrow crab also seeks refuge by withdrawing into a protective shield of another animal. In this case the spines of a juvenile sea urchin provide a shelter under which the crab crawls. The hard shell of the crab's exoskeleton provides some protection for the tiny animal, but not enough to ensure its survival. But the formidable spines that are part of the sea urchin's armor also aid the arrow crab.

Damselfish. *The small size of these apparently defenseless damselfish (above) enables them to live among protective coral branches.*

Spiny refuge. *An arrow crab (right) seeks protection among the sea urchin's spines.*

Covered with Sand

Many bottom-dwelling animals, like this angel shark and the partially buried ocipod crab, hide on the bottom to keep from being attacked. The angel shark, like the skates and rays, has a flattened body and broad pectoral fins. On the bottom, the shark flutters its pectorals to stir up sediments, which settle back on top of it and help to camouflage it. The sand breaks up the shark's outline, making it difficult for prey

*An **angel shark** hides on the bottom, covered by a fine layer of bottom sediment. Note how well the brownish color of the shark's dorsal side blends with the color of the sand.*

and predators to spy it. Hiding in this manner is carried to the extreme by members of the wrasse family. A number of these slender little fish are able to burrow in the sand, headfirst, and completely bury themselves when danger approaches. Some are reported to sleep through the night covered with sand.

92

This crab can quickly dig a burrow in the sandy bottom to conceal itself. It sifts the sand up and over its shell-covered body until only its eyes, situated at the tips of two stalks, protrude. Then it can watch what goes on around it without exposing itself to predators. The light-colored crab blends in especially well with the bottom, and until it moves out of its hiding place, it is extremely difficult to see. Relatives of this crab live just at the water's edge where waves sweep back and forth. With their antennae they filter plankton from the water as the waves recede. Below the sand they are virtually invisible. Their prominent antennae, however, give them away to birds searching for bits of food and fishermen in quest of bait. These telltale antennae lead to the demise of many individual sand crabs.

*This **crab** is an expert at excavation and can dig a burrow deep enough to conceal all but its eyes in a matter of seconds.*

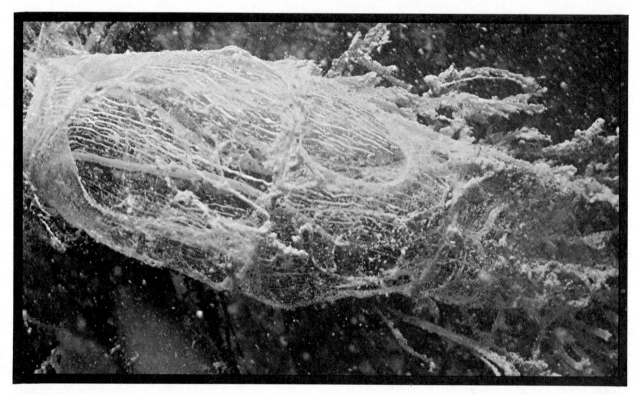

This **Navanax** *(above), found off the coast of California, is nearly hidden beneath strings of mucus it has secreted.*

Navanax cocoon. *When it has no further need of its protective cocoon (below), the Navanax breaks open one end and eases out.*

Mucous Shields

The brightly colored sea slug (*Navanax*) is relatively safe from predation. It seems that only others of its own kind find it palatable. Perhaps one reason why predators leave it alone is the yellow fluid it gives off when disturbed. Another reason may be its odor, which is unpleasant to us and may also offend fish and other animals. The parent *Navanax* imbeds its eggs in mucous coating to discourage predation. Adults also use mucus to form cocoons for their own protection.

Parrotfish are brightly colored residents of the coral reefs found in warm tropical seas. By day these fish graze on the reefs, biting off chunks of coral which they eat for the algae on and in them. By night some species sleep in a mucous envelope. This covering

Sleeping protected. Two queen parrotfish are caught napping in a crevice in the coral reef, surrounded by protective envelopes of mucus. Special glands in the fish's skin secrete the delicate membrane which looks like a cellophane packing.

completely surrounds the fish and may help protect it from its enemies. It may take the parrotfish as long as 30 minutes to secrete its covering. Some scientists think that the cocoon may act as a barrier to prevent the fish's odor from attracting a predator.

> "Some parrotfish sleep in a mucous envelope, which may take as long as 30 minutes to secrete. Although the covering looks fragile, it may take the fish nearly as long to break out of it as it took to build it."

Tunnels and Burrows

A blenny finds a hole in the substrate and simply moves in and adopts the territory around it as its own. The hole may be a crevice in coral, a tin can without a top, or even a broken bottle or other debris. Some blennies burrow into soft sand. Once it has staked out a claim to a piece of real estate, the blenny remains close to its home. It often rests in front of the opening, or just inside it, carefully watching for approaching predators, prey, or trespassers of its own species.

Garden eels dig into the sandy bottom with their tails, excavating long winding tunnels in which they hide at the first sign of danger. When they are not frightened, they reach partly out of the tunnel to feed on passing planktonic animals. They look like a garden of flexible question marks as they wave to and fro, facing the gentle, passing current.

On a trip to the Maldive Islands in the Indian Ocean, the crew of the *Calypso* observed these mysterious creatures. Areas of the bottom were carpeted with garden eels; but when a diver approached within 15 or 20 feet of them, the eels ducked into their burrows and were gone in the blink of an eye.

When the divers tried to remove one from its burrow to observe it, they discovered that the eels plug the entrance to their burrows with a mucus secreted from their skin. Finally some were captured and two were put into an aquarium on the bottom. Soon wrasses and triggerfish swarmed around the aquarium, trying to get at the helpless eels.

Garden eels. *Above, garden eels in the sandy tunnels they dig with their tails. They withdraw at the slightest hint of danger.*

Blenny. *At right, a blenny peers out from its adopted home.*

96

Chapter VI. Offensive Defenses

The drive for survival, which includes getting food and repelling predators, shapes defense and attack systems in most living things.

Poison is one of the defenses developed by a number of sea creatures. It may be administered by teeth, spines, beaks, or barbs. A

> "Man is not as well equipped physically for survival as are many other animals. His modern offensive and defensive devices are all inspired by nature: smokescreens, tear gas, poisons, knives and scythes, hammers and battering rams, stun guns and electric chairs, barbwire and burglar alarms, guns and slingshots. But with the invention of the nuclear bomb, man may have dissociated himself from nature."

few animals have electrical properties they use to stun prey or predator. Many crustaceans and some other animals have pinching claws to capture, crush, and rip food or to defend themselves against each other or other predators.

Whether poison is or is not involved, stabbing alone with sharp spines is often enough to discourage many predators from pursuit. Biting with teeth is one of the commonest defenses, especially among vertebrates. Some bite and hold, some slash as they bite, and some bite repeatedly; all cause blood to flow, thus damaging tissues and vital organs. Larger animals may use brute strength to club or ram an opponent. Sounds are some-

times used to frighten attackers away as much as to paralyze with fear a potential meal.

Some sea creatures change their shape to appear too large for the predator. And conversely a few can accommodate prey bigger than they are through expandable jaws and digestive tracts.

Man is not as well equipped physically for his own survival as are many other animals. With no claws, no fangs, a soft skin, and a modest running speed, he had to face constant hostility from the world and could only survive by fighting nature with all the tricks his brain and his hands enabled him to develop. Recently his conquest of energy sources and subsequent technology brought about the diversification of the weapons in his arsenal: they range from smokescreens, tear gas, and Mace to injections of poisons.

Modern man's devices for attack and defense seem more elaborate than animal systems. In reality, however, they are all inspired by nature, with the erratic exception of the nuclear bomb. One wonders if with this invention man has not dissociated himself from nature forever.

White tails. A hundred years ago, when man went out in small boats to harpoon whales, they feared the havoc that could be wrought by the tail of the whale. The flukes of the great whales could shatter a whaling boat. Whalers gave a name—"bobtailing" —to the whale's noisy habit of slapping their flukes on the water. In this picture we can see a whale's flukes flare upward as the whale begins to sound. Whales have been observed forming a protective circle around a wounded member of their pod. With their heads facing the injured whale and their flukes facing outward, they beat the sea furiously with their flukes to frighten their only traditional foes, orcas and white sharks. When used against predatory man, however, this usually effective defense backfires. Holding in place as they beat the water, the whales were shot by men from point-blank range.

A Spiny Defense

The numerous spines on this sculpin's back are an effective defense just as barbwire keeps intruders from a man's domain. In addition, the sculpin has spines on its cheeks, in its pectoral and pelvic fins, and in some species on its head. Any attempt against such a well-armed fish is disastrous for most predators. Mottled color and ragged appearance further help the sculpin evade trouble. But these defenses come into play only when the sculpin becomes afraid of being attacked. As a threat, the sculpin often spreads its broad, fanlike pectorals.

> "The sculpin has spines on its back, its cheeks, in its pectoral and pelvic fins, and in some species on its head. Any attempt against such a fish is disastrous for most predators."

Spines and Shape

Armed with venomous spines on its back, the scorpionfish is formidably defended. Since its ragged shape helps it blend into its usual habitat among rocky reefs, it is nearly invulnerable. Most members of the scorpionfish family have 12 or 13 spiny rays in their dorsal fins and most have venom in or at the base of these rays. When a living creature confronts it, the scorpionfish arches its back suddenly, erecting the spines.

Torpedo ray. *Lacking a stinging apparatus, the torpedo ray relies on its ability to produce an electric shock to stun its prey and defend itself.*

Electric Shock

The torpedo ray, one of the electric rays, uses its unusual electric generator as an offense and a defense. Being a relatively sluggish swimmer, the torpedo could not capture its prey without its electric-shocking ability. It generates electricity in two large organs on either side of the head; these organs are made up of many hexagonal cells of modified muscle tissue. Large torpedoes can produce electric discharges of short duration, high intensity, and low voltage (50-60 volts). When repeated, the voltage decreases with fatigue; organic batteries take some time to be chemically refueled.

Bat ray. The stinging spine at the end of the bat ray's tail is used as a defensive mechanism. But some basses feed on stingrays with apparent immunity.

Barbed Whiplike Tails

The bat ray has a short, stout, barbed stinging spine at the base of its long, whiplike tail. It is a strictly defensive weapon. Although the bat ray swims, often it stabs an attacker by a flail of its tail in much the same manner as the bottom-dwelling rays do.

The result is devastating to most animals hit by the barbed stinger. It is a serious but rarely fatal injury to man. A person hit by a bat ray's barb feels intense pain accompanied by swelling, redness, and tenderness of the affected part plus swelling of the lymph gland serving that area. The pain and other symptoms may last several days.

103

Borrowed Poison

The delicate flowerlike sea anemones (center) and their relatives, the hydroids, all of which are really marine animals related to the jellyfish and corals, have stinging cells or nematocysts in their tentacles to stun their prey or protect them against predators. Some predators, however, like the nudibranchs (left and right) with white-tipped naked gills on their backs, are unaffected by stinging cells. In fact they eat them without destroying them and use them for protection. These beautiful molluscs have the amazing ability to prevent the discharge of the stinging cells as they are consumed, as they pass into the digestive tract into and up a special canal, and are finally incorporated into their cerata or gills. It is known that the stinging cells have a projecting trigger that responds to touch and it seems logical that feeding actions would stimulate it. One possible explanation to the fact that they are not stimulated is that the nudibranch may inhibit the discharge of the nematocyst chemically, in much the same

way the clownfish gains immunity from the anemone. When predators attack nudibranchs, the ill effects of the stinging cells are passed on. Just how this is accomplished is not clear yet, but it has been proposed that cells discharge only in response to certain types of pressure or contact.

Another defensive adaptation of these naked snails is their coloration. Nudibranchs are among the most vividly colored animals in the sea possessing vibrant orange, blue, purple, yellow, and red pigments. No preda-

tor could mistake any of them for a conventional prey. This is precisely the object of this eye-catching publicity for any animal that recognizes them (and it is assumed they do, having learned or having the ability to react instinctively) will not want to eat such a stinging snail. In contrast to camouflage, in other words, they are brightly colored, and advertise themselves as unpalatable.

Stunning beauty. Sea anemones and hydroids stun everything that comes in contact with them—except the nudibranchs on the left and right in the photograph below.

The Best Defense Is a Good Offense

Many of the more than 5000 varieties of nudibranchs display graceful crowns of soft respiratory organs (cerata) on their backs. Delicate, soft-bodied, and brightly colored, these delightful creatures would appear to be easy picking for any hungry predator. But these bizarre crawlers seem to flourish. The little animals secrete a mucus, which smells unpleasant to man and perhaps makes them unappetizing to fish and other predators.

When provoked, some species exude a strong acid and others discharge a powerful poison. The secretion of one species is reported to be fatal to fish and crustaceans. Another species has a specialized acid gland from which it releases a slimy sour secretion containing sulfuric acid. These are formidable reasons why nudibranchs make very little contribution to the ocean's food chain.

Some of the nudibranchs enjoy even further

Flashy warning. These closely related southern California nudibranchs conspicuously display themselves, possibly warning they are not good to eat.

advantage. They are able to swim. They propel themselves by bending their bodies from side to side with head and tail almost touching. They can also beat the water with their cerata for additional speed. By moving up into the water, they get away from any possible danger from an obstinate foe. Others are able to cast off parts of their bodies when they are under attack and get away. Later, these parts are regenerated.

Despite all these varied and formidable defenses, nudibranchs are not immune to all predators. Some of them fall victim to parasitic worms and copepods and are eaten by some starfish.

These few predators aside, it seems that nudibranchs have enough going for them to be quite safe in their gaudy dress.

Poison Flesh

As if its covering of sharp spines were not enough to deter predators from the porcupinefish, many possess an additional defense: they are, when eaten, extremely poisonous. About one-half of the puffers and porcupinefish are reported to be toxic. The poisonous chemical is most concentrated in the ovaries or testes, the liver, and the intestine. The muscles of the body are not poisonous but the skin is. The Japanese eat these fish, but those who prepare these special dishes must undergo a period of training and must be licensed by the government. The emperor is not allowed to eat pufferfish.

The symptoms of poisoning in mammals are an initial lethargy, weakness, and loss of coordination. Gradually these problems of the nervous and muscular systems increase until death results. The action of the toxin is to disrupt the transmission of nerve impulses to muscles and certain centers of the brain.

Poison-Fang Blenny

This Pacific Ocean blenny has poisonous fangs, canine teeth in its lower jaw, that deter most predators. Groupers, large-mouthed fish that swallow their prey alive, have been seen spitting out a poison-fang blenny immediately after ingesting it. The blenny had probably bitten and poisoned the grouper. The blenny's bright coloration probably serves as a warning to would-be predators that the blenny is a bad risk. As with many poisonous animals the blenny has no fear of others larger than itself and even acts aggressively towards them. It can be assumed that this aggressive action has become a kind of behavioral warning to the fish's enemies. It threatens any intruder into its territory, including divers, with a series of short, jumplike strokes in the intruder's direction. There are, incidentally, two species of nonpoisonous blennies which look very much like the poison-fang blenny, and they probably benefit greatly from this mimicry.

Suction-Cup Feet

A sea star, wrapped around a blue mussel, pulls the shellfish's valves, or shells, apart by sticking its tube feet on either side and pulling in opposite directions. There has been considerable debate over the mechanism by which the sea star opens the valves of a mussel, clam, or oyster. Some scientists thought that the starfish forcefully pulled the shells apart. But it has recently been found that the closure of the mussel's protective shell is not a perfect seal. The starfish inserts its stomach into the mussel's valves and secretes its digestive enzymes. As the starfish digests the mussel and the mussel's muscles weaken, the shells gradually spread farther and farther apart. Ultimately the valves gape and the starfish can completely consume the defenseless mussel.

The attack. The suction cup feet of this attacking sea star allow it to firmly grasp the shells of its victim and expose its flesh.

Crown-of-Thorns

These tough-bodied sea stars crawl relentlessly over coral reefs and devour the living coral animals by everting their stomachs over them and digesting them on the spot. Unlike most sea stars, the crown-of-thorns has sharp, venom-bearing spines, which protect it against most predators. The spines consist of calcium carbonate and organic material and are held erect by muscles. Glandular tissue is reported to be contained within the spines themselves and can secrete a toxin into the water or into the tissue of a would-be predator. When a person is punctured by these spines, he experiences pain, redness, swelling, some muscle paralysis, nausea, and possibly vomiting. In addition to the venomous spines the crown-of-thorns has another poisonous chemical in the skin covering its entire body. The major enemy of the crown-of-thorns is another crawler, the triton snail, itself a large animal.

The venomous spines of the crown-of-thorns starfish are an effective deterrent to all animals except one—the triton snail, its only known predator.

Sea Urchin's Defenses

Sharp spines that sometimes transmit venom to the victim are the first line of defense of most sea urchins, which are relatives of the sea stars. Besides the spines, sea urchins have pedicellariae, three-part jaws on stalks. In venom-bearing species, the pedicellariae transmit the poison. Yet another defense sea urchins have is the shell, or test, in which spines and pedicellariae originate. Like the spines, the test is made of calcium carbonate and is hard but brittle. When attacked, a sea urchin brings its spines into play, waving them about and aiming their sharp tips at its attacker.

Some sea urchins, like this white-tipped one, have added an additional defense. As we see here, they are able to pick up debris from the bottom they dwell on; moving these loose fragments from one spine tip to the next, they finally place the debris on top of themselves. One reason for adopting this cover may be for protection from sunlight. The debris may also serve as camouflage. Thus,

an urchin living on tropical turtle-grass flats may be covered with bits of turtle grass. If it lies on a pebble-strewn bottom, as shown here, it will disguise itself with pebbles.

Sea urchins. *The sea urchins in the photographs below and to the right are perfectly fit to take care of themselves. Some species have movable spines, and in addition some have a structure at the base of these spines that give off a poison. Other species, like the one below, are capable of covering themselves as protection from harsh sunlight, and as a means of camouflage. Many sea urchins live on rocks in shallow water, and those that live deeper down tend to live in groups. They range in size from one-half inch to ten inches.*

Pistol shrimp. The pistol shrimp above uses its large claw to create a snapping sound which stuns prey and startles predators.

Creating a Loud Noise

Loud sounds frighten many animals, and some sea creatures use noises as a defense. For example, pistol shrimp use their large claws to produce the sharp snapping sound that gives them their name. The sound is so loud it can be heard easily by beachcombers or by oyster fishermen, as these snapping shrimp also live under clumps of oysters. But its defense also works against the shrimp; it can give away its presence when it might otherwise be unnoticed. Mediterranean fishermen often find distant rocky shallows out of sight of land by listening to the snapping shrimp.

Crab. Protected by its exoskeleton, this tropical crab remains unconcerned by the presence of an underwater photographer.

King Crab

The hard outer shell, or exoskeleton, of this tropical king crab is its principal defense. Its armor, coupled with its large size, makes it a tough consumer product. The pincers on the crab's long, spidery legs can also give an attacker a good nip. Its cousin, the northern king crab of Alaska, grows to 20 pounds and may be five feet wide; its size alone discourages many predators. The marine growth that accumulates on its 10-to 12-inch body shells, or carapaces, acts as camouflage. When approached, most crabs thrust their arms upward, brandishing open claws in a threatening manner.

Chapter VII. Fighting for Territory and Sex

When combat takes place between members of the same species, it is almost always the result of competition for territory or a mate. Initially, one animal may assume control of a territory by driving off another occupant or by just moving into an unoccupied area. If challenged for control of its home ground, the animal will defend it vigorously. Similarly, two males competing for the favors of a single female may engage in ferocious ritualistic combat for the prospective mate. In both cases, possession is the motivation.

Open-ocean species are generally not territorial, but inshore animals frequently are. To ensure their survival both as individuals and as a species, it is important for territorial animals to own a home unencroached upon by others. Its territory gives an animal a place to hide from predators. It also provides a place to feed without the competition that could lead to overgrazing and subsequent shortage of food. Ownership of territory also plays a major role in the sex lives of some animals. The territory provides a place to mate and lay eggs or give birth to offspring.

Fighting for the right to mate is to the advantage of the whole species, since the strong characteristics of the victor are passed along to the offspring. The loser, usually not as fit as the winner, is prevented from passing along his weaker traits.

Although combat for sex and territory seems dangerous to the participants, injuries are usually not very serious and rarely fatal. Animals generally do not fight among themselves the same way they fight an enemy. In fighting members of their own species, they meet in ritualized combat, involving the use of threat displays and other noninjurious means to defeat a challenger. The weapons they use against predators and prey (such as teeth, claws, and poisons) are rarely or partly used against each other. These weapons are intended for killing, and killing is not in the interest of the species. So they resort to violent, but usually harmless, jousting. Though there are notorious exceptions, such as deadly fights of octopus for a shelter, a growl, grunt, or squeak may be

> "The size of participants is not indicative of their ferocity. The smallest creatures fight with at least the same resolve as the largest."

enough to drive an opponent away. Some fish beat at each other with their tail fins. The pressure wave they set up is indicative of the strength of the fish. Often, though, more subtle means are employed to show superiority. Sometimes opening the mouth wide is enough of a threat to frighten off a challenger. Color changes, ritual movements, or flaring of the fins may do the job too. The size of the participants is important in the intimidation game, but is not indicative of their ferocity. The smallest creatures fight with the same resolve as the largest until one finally emerges victorious and allows the loser to leave and fight another year.

The jawfish is extremely territorial and does not like to wander far from the safety of its burrow. At the approach of another jawfish, it quickly darts out of its burrow and rushes toward the intruder as though to strike it. But it seldom hits its adversary; it returns to its burrow nearly as fast as it left, diving back inside when the confrontation is over. This fish is named for its highly eversible jaws, which it opens wide to threaten an intruder. An open mouth probably makes the fish appear larger than it is. As in most battles between animals of the same species, neither is harmed. Locking jaws is the most violent act of warring jawfish.

Mantis shrimp. *This animal normally fights to protect its territory, spreading its armored tail over the entrance to its lair. This usually deters any challengers.*

Physical Combat

Most animals end a battle with others of their own kind before coming to blows. Mantis shrimp live in cavities in rocks, corals, sponges, or in empty shells, while some larger species prefer to drill their own cylindrical home in the sediments. Their fighting is generally over occupancy of a cavity. Combat between mantis shrimps follows a strict ritual. The challenger approaches boldly, head high, antennae in motion, eyes fixed on the goal. The occupying mantis spreads its armored tail to block the entrance to the cavity. This action is often enough to discourage the intruder.

118

Eel blind. *To study and photograph a colony of garden eels in the Red Sea, a small blind was erected. Divers behind the structure did not disturb the animals.*

Garden Eels

Colonies of garden eels live in individual burrows dug in the ocean floor. So timid of exposure is the garden eel that none has ever been seen wholly extended by uncamouflaged divers. There appears to be a rigid social order in garden eel colonies, with dominant males, harems, and firm property lines. A challenger from an adjoining territory is met with a ritual display. Stretching from the burrow, the eel performs snakelike undulations, ripples its dorsal fin, and turns its profile and then its back to the attacker. If the challenger persists, the eels will square off, snout to snout, and strike at each other.

Lobsters in Combat

Atlantic lobsters in combat use their large pincers. Here a pair face off and, unlike many other creatures in the sea, actually do bodily harm to each other. In this fight for territory, one lobster has succeeded in literally disarming the other. The loser could be killed and eaten by the cannibalistic opponent. Or it could limp off to regenerate a new claw to replace the old one. Regeneration to full size may take several molts over a two-year period. Atlantic lobsters usually have their hard exoskeleton to protect them against other lobsters or other marauders, and if necessary, they can shed their claws to escape.

Lobsters are most vulnerable immediately after they have molted. Males and females alike are subject to harassment and dismemberment by their own kind and others when their new shells are no stronger than wet paper. In the several days after they shed their old shell and before their new ones have hardened, they spend much of their

time in hiding. During this time they have their greatest need for calcium carbonate, which provides the hard substance of the shell. The lobster's new shell begins forming under the old one and is principally composed of soft proteinaceous material. This shell swells when the old one is shed and provides the animal a little room for growth. The action of seawater upon the protein-aceous material and the addition of calcium carbonate to it cause the shell to harden. The lobster's need and craving for lime or

Battling lobsters. *The loser of this battle may lose anything from a claw to its life, depending on the cannibalistic nature of the victor.*

calcium carbonate is great, and it is not un-common for a lobster to consume its cast-off exoskeleton immediately after shedding it. It is, in part, the calcium carbonate deficiency that has contributed to the difficulty of lobster mariculture. Many attempts have been made to raise lobsters, but fighting is a problem in crowded tanks.

Fearless Guardians

Wrecks and coral or rock reefs are favorite haunts of groupers. When they find a cavern that suits their needs, they stake it out and call it home. They are extremely defensive of their home territory and do not allow passersby to trespass, regardless of their size. This pair was guarding their territory when its boundary was violated by the photographer. Feeling threatened, the male was openly aggressive and rushed the diver. It hovered brazenly before the camera, its mouth opened wide in characteristic threat display, until the diver backed down. Its partner hovered beneath and behind it, being slightly more cautious.

The brilliant-orange garibaldi lives on the rocky shorelines of southern California and Baja California. It lives among the great kelp forests there and defends its territory year round. During July and August when the females are ready to spawn, the males become more defensive than normal about their home site. They fastidiously clean around a clump of red algae to prepare it for the female to deposit her eggs and courageously attack any animal who moves in with intent of taking it away.

Anyone who has observed this fish underwater, whether as a diver or from a glass-bottom boat, has surely marveled at its brilliant color, particularly when compared with its usually drab neighbors. Obviously, camouflage is not important to this fish. It has been postulated that such vivid colors may serve as a means of identification to others, which would supposedly deter them from approaching this pugnacious fish.

*Pictured above are **two groupers,** whose home territory is under threat. Being highly territorial animals, groupers fear little. By opening its mouth wide, the larger grouper displays its characteristic threat.*

*This **garibaldi** hovers alertly near its nest, ready to defend its territory against intruders, no matter what their size. Their aggressive displays often include the production of loud grunts or thumps.*

Kissing Snappers

As these gray snappers square off in ritual-ized combat, they seem almost ready to kiss. A confrontation generally does not go be-yond this stage. For many reef fish, disputes arise over the control of territory, but con-clusions are difficult to draw, as fish behavior is disturbed by the intrusion of an observer, and fish do not conduct themselves normally when they are in captivity. Whether or not their opened mouths are intended to make them appear larger has not been confirmed for these fish.

Such bloodless confrontations are purely in-stinctive reactions, not a form of learned behavior. To prove this, experiments have been undertaken in which two animals were reared separately and in isolation from others of their species. When the isolated fish were introduced to another of the same species, they exhibited the inherent ritual-ized behavior seen in their normally reared relatives.

Aggressive behavior towards a member of another species may differ drastically from that towards a member of the same species. In attacking a nonrelated intruder a fish may actually use its teeth or whatever defenses are available to do physical harm. This difference shows that the nondestructive attacks or ritualized threats are really a form of communication. There are many advantages to having rules that enable opponents to threaten, attack, and win or lose.

Threat display. *Despite looks to the contrary, these snappers are not ready to kiss. Their ritualized pose is a threat to one another, probably over territorial rights.*

The victor could remain a productive individual unharmed by battle and the loser could eventually succeed as others became weak or died. The population as a whole would be able to keep its numbers up and remain viable. The field of international relations is an obvious analogy.

Fighting for Social Status

Battles between elephant seal bulls often occur during mating season and are usually intended to determine the social order within the herd. Young bulls, not yet sexually mature, engage in mock bouts, rehearsing for adult combat. Before body contact is made, the bulls threaten each other with an inflated snout, a raised stance, abrupt aggressive movements and alarmingly boisterous bellows. When one of the opponents is obviously weaker, it will retreat before blood is shed. Only one in 60 of these confrontations gets beyond the threat stage. But if the males are comparable in size, weight, and aggressiveness, threats are ignored, the animals square off, and the fighting begins.

Nearly all fights occur on land, but the combatants may move off the beach and into the water. Land-based fights are generally short, lasting five minutes or less. But the seals sometimes battle for 45 minutes.

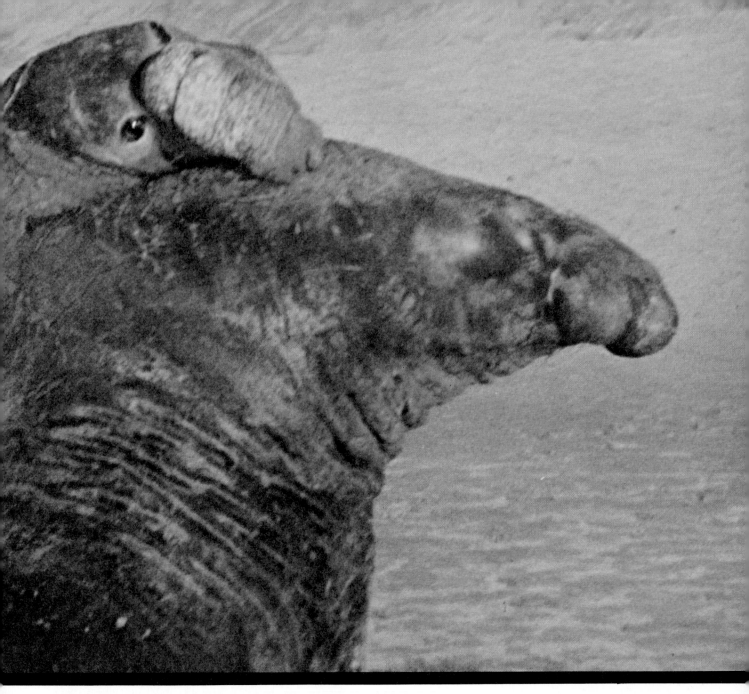

The two behemoths stand chest to chest, feinting and faking, waiting for an opening to fight. Finally, with a fast and powerful blow, one strikes at the neck of its opponent. The attacker's head slashes downward and its sharp teeth rake the opponent's flesh.

When one has had enough, it backs away, conceding defeat. The loser has not necessarily received the worst of the battle, but for some reason it chooses not to hold its ground. The bleeding is profuse and the

Elephant seals. In the bloody scene above, two elephant seal bulls are locked in combat. Battles of this sort most often occur during the mating season, usually on land. The fighting is brutal but short-lived, and although the wounds inflicted are deep, elephant seals heal rapidly.

wounds are deep, but they heal quickly. Fortunately the neck and chest of the elephant seal can stand up to this rugged treatment and have a horny layer of tissue to help protect the animals from serious injury.

Chapter VIII. Ancient Animosities

Relationships among people and other animals take many forms. Some species get along with each other. Other creatures have so little to do with one another's lives that theirs must be termed a nonrelationship. There is, for instance, the case of the anemone (*Epiactis*) and the turban snail (*Tegula*). They simply do not care about each other. They do not look to each other for food or shelter; nor do they crave the same food or shelter. The anemone eats microscopic

> "There are a few species which seem to have a natural animosity toward other species. These antagonisms give the appearance of being fraught with deep emotions of fear and hate."

plankton, and the snail's diet consists mostly of fine algae. Neither animal is interested in the kelp it rests on. There are many relationships like this one in the sea in which two or more animals can live in close proximity with one having no bearing on the lives of the others.

Some animals even help each other. But there are a few species which seem to have a natural animosity toward other particular species. Such are cats and dogs, cobras and mongooses, and in the ocean, sperm whale and giant squid. These antagonisms give the appearance of being fraught with deep emotions of fear and hate. Whether they actually are is highly doubtful.

Why and how do these traditional animosities arise between species? When two species don't get along, their aggressive behavior could be emotional or calculated. Perhaps hundreds of thousands of generations ago the members of two species competed for the

same habitat, territory, food, or ecological niche. In competing, they may have resorted to combat. And perhaps this combativeness has continued as one species faces the other, even many generations after one has adapted to a different niche and the two are no longer competitive. Octopus-lobster, moray-octopus, and lobster-moray are three couples with such built-in animosities. Mediterranean fishermen tell stories of traps that they pull back to their boats, sometimes containing an octopus, a lobster, and a moray eel: the three retreat to the three corners of the trap as far from the others as possible, because they know the first one to attack will be immediately killed by the third party.

Hard-shelled snails have enemies that are able to overcome their rugged defenses. At right a large tulip snail is devouring a small helmet snail. The larger animal works on the trapdoor (operculum) of the smaller, gains entry to the shell, and consumes the soft flesh. Many snails are active predators on creatures they are able to catch in a slow-motion race; others consume marine algae.

Anemone and turban snail. The animals below inhabit the same environment, but live by different modes and pay no attention to each other.

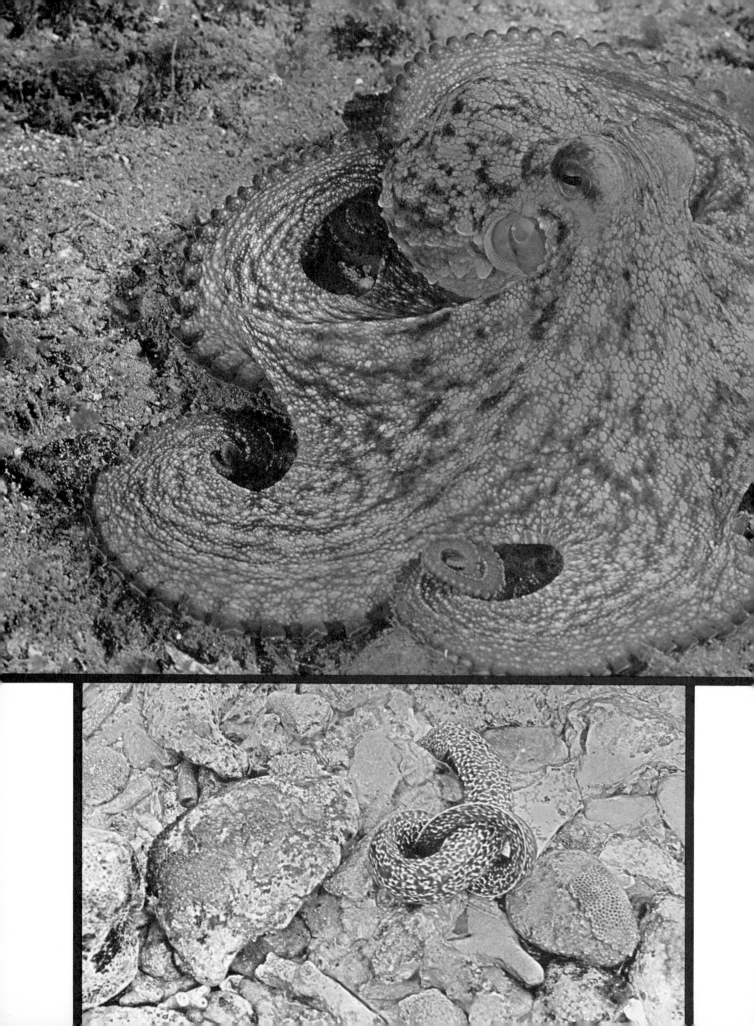

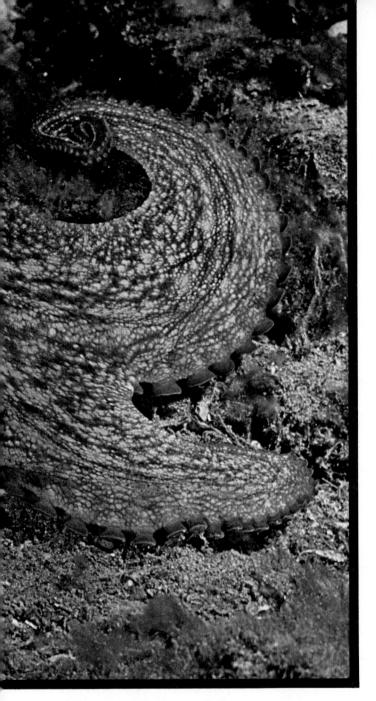

reacts immediately by wrapping its tentacles around the moray's head. The moray then ties its body into an overhand knot by looping its tail around and over itself. The eel withdraws through the knot sliding out of the octopus's grip while retaining a jawhold on it. Quickly, then, the moray can start gulping his prey.

The viselike grip of a moray, reported by divers who have been bitten, is almost impossible to break until the eel decides to give up. The teeth of morays are particularly adapted for grasping prey. Projecting from both the upper and lower jaws, and even the

> "The octopus reacts by wrapping its tentacles around the moray's head ... The moray ties its body into an overhand knot ... Then the eel withdraws through the knot, sliding out of the octopus's grip while retaining a jawhold on it."

roof of the mouth, are sharp, slender, needlelike teeth that are essential for any fish which must hang on to an octopus.

If the moray is not successful in its first couple of attacks on an octopus, it will very likely go without a meal. This is because the ink of the octopus can act in two ways to assist its retreat. Film footage has revealed that the first advantage to the octopus is that a cloud of ink presents a false object for the eel to strike at. The second asset lies in the fact that chemicals in the ink temporarily deaden the sense of smell in the eel.

Moray Eel and Octopus

When octopus meets moray, a struggle to the death ensues between these mortal enemies. Morays usually prey on octopods at night. When a moray smells an octopus, the eel slips in and out of holes throughout the area, zeroing in on its quarry largely by scent (the moray's eyesight is not particularly useful when feeding at night). When the eel finds the octopus, it clamps its jaws on it and the struggle begins. The octopus

An octopus (above) watches warily; it has changed its color to match its background and thus escape detection.

Making a knot. A spotted moray on a rocky beach (below) ties itself into a knot. This is similar to its reaction to an octopus's grip.

131

Penguins and Skuas

In the southernmost parts of the Southern Hemisphere, where the Adélie penguins pictured here normally nest, there is a species of predatory oceanic bird called the skua. These birds have characteristics of both the gull and the hawk. They are similar to the gull in size, habitat, and food preferences. And they have the hooked beak, the predatory habits, and the general appearance of the hawk. Skuas are related to gulls and to a group of seabirds called jaegers, a word which means "hunter" in German. The term could well be applied to the skua as well. Twice a year, the skuas prey especially on the Adélie penguin. Normally the skuas are fish eaters, and they dive into the water to capture fish. But when the Adélie penguins lay their eggs, skuas returning from their fishing waters often seize and devour some of the eggs that are unattended. Not long after, when the chicks have hatched, skuas again prey on the weakest of the baby birds, ripping the helpless chicks with their hooked beaks. The baby penguins are defended by adults who very

courageously and very successfully chase the skuas. The skuas are frightfully obstinate; they remain nearby waiting for any young left unattended.

In the northern parts of the Northern Hemisphere, where skuas also occur, these hunters prey on fish and rarely on the arctic birds, such as gulls and terns that may come near them.

Skuas seem to have an underserved reputation for being a vicious predator. They are a predator no different than any other and

Adélie penguins. Every year, as they settle on land to breed, some of their eggs and chicks are preyed upon by a species of gull, the skua.

actually, in the long run, benefit the penguin population as a whole. Recent observations prove that most penguin eggs taken by the skuas were either unfertilized or dead before hatching and that their role in the rookery is more that of a scavenger than of a predator. They also might remove the behaviorally or physically less fit, indirectly making the species stronger.

Diver, Sea Star, and Octopus

After being dropped in front of an octopus, this sea star rights itself and attempts to move off. The octopus probably will not feed on the starfish. The withdrawn appearance of the octopus would indicate that it may be more concerned with the presence of the diver than that of the starfish. One must be cautious about making conclusions on the natural behavior of animals when they are under influence of a bright strobe flashing and a diver noisely blowing bubbles.

A / Righting. *The flexible arms of the starfish reach over to contact the bottom. Righting occurs in a surprisingly short period of time.*

B / Octopus passively watches. *The octopus has withdrawn into its den at the approach of the diver.*

C / The starfish retreats. *Starfish also possess suction cups under tube feet which propel it slowly away.*

D / A curious blenny. *A blenny darts about in front of the camera. It may be curious, or it could be threatening the diver, who has intruded on its territory.*

E / The octopus retreats. *Eventually the octopus has had enough of the diver and his bright light and passes over the starfish in its retreat.*

▲ A   ▼ B

Urchins and Their Enemies

Despite their impressive defensive equipment, sea urchins have many enemies. The most dramatic example of sea urchin predation is found in the feeding behavior of the queen triggerfish. The *Diadema* sea urchin has an interesting defense system which must be overcome by the triggerfish. When any point on the urchin is touched, it directs all its long thin, highly mobile spines toward that point of contact. This makes it almost impossible for a predator to penetrate the spiny coat to attack the more vulnerable shell. The triggerfish cannot attack the dorsal surface and must instead focus its attention to the ventral or undersurface where the spines are very short. The problem for the fish is how to get to this vulnerable side. The solution lies in the queen triggerfish's ability to grasp the longest spines of the urchin in its teeth, swim up off the bottom, then let go. As the hold is released, the urchin simply falls to the bottom, usually ventral side down. The triggerfish is persistent, however, and continues until the urchin happens to land on its side. As soon as this occurs, the triggerfish is able to whirl around and attack the urchin's vulnerable underside. The short spines are not enough of a deterrent to prevent the sharp protruding teeth of the triggerfish from biting into the mouth of the urchin and eventually through the shell into its insides. It is difficult to imagine how a supposedly unintelligent fish was able to develop this behavior.

Another interesting predator is the sea otter which also uses its brain to overcome the formidable defenses of the urchin. Along the coast of California, otters have been observed floating on their backs breaking open the spiny test of the urchin, using rocks as tools for pounding.

Another less dramatic predator upon the

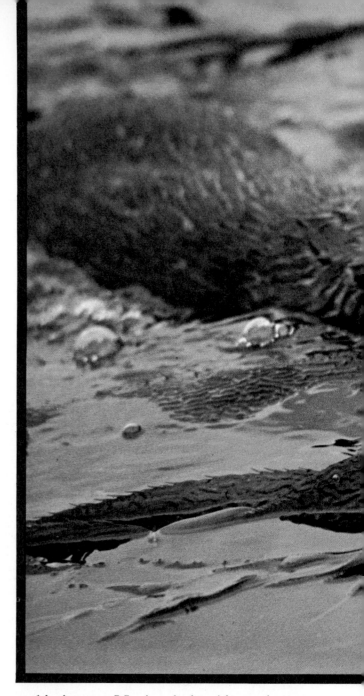

urchin is man. Man's relationship to the urchin is paradoxical. Some people spend considerable effort seeking the urchin as a gastronomic delicacy, while other people detest the urchin and wantonly kill it with chemicals and hammers. Many Europeans (particularly the Italians and French) and the Japanese savor urchin gonads.

A sea otter (above) *swims up to purple sea urchin, offered on paddle next to kelp, to eat it.*

Black sea perch (right) *feed on remains of sea urchin lying split open on ocean floor.*

Belugas v. Orcas

These handsome belugas, or white whales, of the Arctic Ocean are as successful as dolphins in their cold realm, but they recognize and fear one old enemy—the orca, sometimes called the "killer whale." Scientists are using this fear to reduce predation by belugas on the salmon of the Canadian Pacific coast. In one experiment conducted in Bristol Bay, Alaska, orca sounds recorded on tape and played back into the river through an appropriate underwater speaker repulsed almost 500 belugas attempting to go up the river to prey upon young salmon. During the three-week period when the salmon are migrating down the river, the white porpoises

> "Belugas recognize and fear one old enemy—the orca. Scientists are using the fear to reduce predations by belugas on the salmon off the Pacific coast of Canada."

138

enter the waters twice daily to feed. During six different times of orca playbacks, the belugas were observed immediately to turn downstream and swim rapidly out to sea. On the seventh trial, the pod of whales separated and swam out, the eighth playback induced them to swim to the far side of the river, and finally, on the tenth session, all whales continued up the river, but on the far side of an adjacent sandbar.

Obviously the belugas eventually learned that the sounds were not from a live orca.

Belugas. *Scientists are now able to capitalize on the beluga whale's fear of the orca. By playing the sounds of the orca into a river, they have repulsed the beluga from feeding on salmon.*

However, the scientists feel that this deterrent will be effective during the short critical periods when the greatest number of salmon migrate in the rivers. It will be interesting to see how many of the whales remember man's clever deception from year to year. It may be that in the near future the beluga population will ignore the orca calls entirely.

Sharks v. Dolphins

In the deadly but careful game of shark v. dolphin, the mammals eventually win. But not always—as here we see a shark devouring the remains of a dolphin. The greater intelligence and vitality of the dolphin gives him the advantage. In great number, sharks always trail packs of dolphin, waiting for a dropout—an ill animal or a young one—to fall behind the rest of its group. Then the shark moves in to devour the dolphin. Actually the sharks are opportunists and never attack in the sense of a tuna stalking sardines. The dolphin is not a vicious warrior either. It surely does not search the seas for sharks simply for the joy of killing, but it can and does frequently get rid of sharks for safety reasons. Turning on the speed, they slam beak-first into the gill area of the soft lower abdomen of the sharks, their most vulnerable spots. The dolphins' beaks—their pointed jaws—are efficient weapons for such blows.

Several marine laboratories are studying the shark-dolphin relationship in the hope of making the dolphin's behavior useful to man. Experiments on lemon sharks and bottle-nosed dolphins show that if given the choice, sharks will avoid dolphins. The researchers have also been training dolphins to make sure that they could be used for shark control. One dolphin has been taught to ward off sharks in captivity on command from a sonic device. The dolphin, on cue, will chase and hit the shark. Soon the scientists will

The unfit. Despite the fact that dolphins usually win in battles against sharks, sharks often wait for an unfit dolphin to drop out of a group so they can devour him.

conduct these experiments in the open sea, hopefully employing dolphins to defend divers from sharks. Someday such trained dolphins may help oceanauts by acting as watchdogs around undersea habitats, or they may police coastal beaches, warding off sharks and protecting swimmers.

A Time for Peace

For living dots and stars and germs
For jelly barrels, feather worms
For buds and bugs and shrimp and larvae
For clams and crabs and anemone

        With sham and poison, bites or gulps

Guns and armor or suction cups
Mean tricks and struggles—sly escapes

        Only to own or eat or mate.

In aimless drift, for pulsing clouds
On ocean floor for humble crowds
Which toil and moil in water winds
There is no break from fear and hunger.

In frantic feasts at dawn and dusk
Wings, fins, and flippers bold and shy
Glance off and back their wakes and gleams

        When fangs and beaks have torn their spoils
        A fragile peace spreads in the Sea.
        Longing for more exotic foe
        The thoroughbred, masters of Space.

Hairy quicksilver shapes that wing out of the brine
Or sound into the dark for iridescent prey.
Entrenched in oblivion denizens of the gloom
Squirt up to brighter realms and rape a beam of sun.

        The Ocean Lords with time to spare
        Display the Spirit of the Sea.

Index

ILLUSTRATIONS AND CHARTS:

Howard Koslow—31, 35.

PHOTO CREDITS:

Chuck Allen—24 (top); Ken Balcomb—99; John Boland—37, 49 (bottom); Tony Chess—88; Jim and Cathy Church—84, 85; Ben Cropp—2-3, 16, 21, 87; David Doubilet—17, 22, 26, 27, 42, 43 (right), 65, 67, 91, 101 (bottom), 119, 120-121, 122; Jack Drafahl, Brooks Institute of Photography—44 (top), 97, 106, 114, 128; Freelance Photographers Guild: Patrick Colin—117, James Dutcher—56-57, 61, 70, Robert B. Evans—49 (top), FPG—72, L. Grigg—89, Jerry Jones—92, 115, Stan Keiser—80-81, 123, Tom Myers—11, 36, 64, Chuck Nicklin—29, 52, 62, 71, 90, 108, 124-125, John Stormont—118, A. B. Trail—48 (bottom), 50, 69, 73 (bottom), Western Marine Laboratory—14, 38-39, 41, 82-83, 110, 137 (bottom); H. Hansen, Aquarium Berlin—109; Edmund Hobson—95; C. Scott Johnson, Naval Undersea Center, San Diego, Calif.—30, 140-141; Holger Knudsen, Marine Biological Laboratory, Helsingor, Denmark—76; Don Lusby, Jr.—138-139; Jack McKenney, Skin Diver Magazine—18; Maltini-Solaini, M. Grimoldi, Rome—102, 111; Charles Mather—19; Richard Murphy—142; Chuck Nicklin—103; Carl Roessler—46, 47, 51, 86, 130 (bottom); George X. Sand—12; Paul R. Saunders, University of Southern California—34-35 (top); John B. Shoup—93, 132-133; Tom Stack & Associates: Douglas Baglin (NHPA)—23, Bill DeCourt—104-105, E. R. Degginger—13, 79, Warren Garst—126-127, Dave LaTouche—129, Tom Myers—96, William M. Stephens—66, 78; Joe Thompson—15; U.S. National Marine Fishery Service—24 (bottom), 25; Don Wobber—77; Ed Zimbelman—5, 68.